MODERN FIREARM SILENCERS

MODERN FIREARM SILENCERS

GREAT DESIGNS, GREAT DESIGNERS

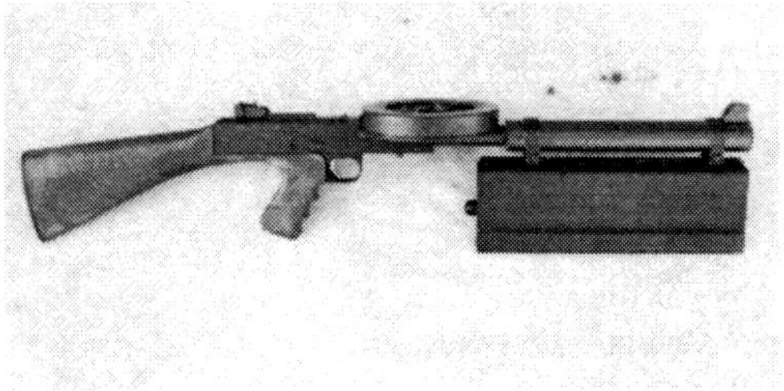

J. DAVID TRUBY

PALADIN PRESS
BOULDER, COLORADO

Also by J. David Truby

Improvised Modern Firearms: Deadly Homemade Weapons
Silencers, Snipers, and Assassins
Zips, Pipes, and Pens: Arsenal of Improvised Weapons

Modern Firearms Silencers:
Great Designs, Great Designers
by J. David Truby

ISBN 10: 0-87364-666-5
ISBN 13: 978-0-87364-666-6
Printed in the United States of America

Published by Paladin Press, a division of
Paladin Enterprises, Inc.
Gunbarrel Tech Center
7077 Winchester Circle
Boulder, Colorado 80301 USA
+1.303.443.7250

Direct inquiries and/or orders to the above address.

Contents

Preface

In 1971, as I was finishing exhaustive and exhausting research for *Silencers, Snipers and Assassins,* I corresponded with Bob Brown's partner in Paladin Press, the businessman who signed the letters and checks. He was only a name on paper then, a signature called P. C. Lund. Today, Peder Lund is one of my closest friends, a man I would literally trust with my life . . . and there ain't too many of those folks, believe me. Peder and I have never had a signed contract—unheard of in the publishing business. We did it all with one handshake. I'm telling you this because I wanted to thank my friend in public for his help and friendship over the years.

It's been just over a decade since my first silencer book was published. My subsequent books on the topic have been mostly updates and war stories that never approached the depth of research of the first book, in my opinion, anyway. This present book does measure up to the original, though. I am very happy with this book and I hope you will be, too.

My "thank you" list would be endless if I included everyone who sent a photo or a clip from a publication, passed along a tip, told me a silencer story, or steered me to some home workshop inventor. I am also leaving out mention of public relations people and others whose paid job is to produce favorable images for their employers, products, clients, etc. I must also leave out the names of a large number of sources for this book—people who cannot have their names publicly acknowledged because of political, social, legal, financial, and, chillingly, even health reasons.

That leaves the following folks who also deserve special recognition for their valuable contributions to this book: Mike Albanese; John E. Bacon; Frank C. Brown; Jonathan Arthur Ciener; Barbara Demato; Philip Dater; Sheree D. Frederick; Jean-Pierre Gillet; Jack Krcma; Jack Robbins; Emilio Santana; Robert Scroggie; William Seymour; Donald Thomas; R. K. Thomas; and Don Walsh.

Also, I can never forget the contributions of Pike Bishop; Dutch Caleb; the Gorch brothers, Lyle and Tector; Freddie Sykes; and Deke Thornton, all of whom I've seen within the week.

J. David Truby
Zihuantanejo, Guerrero
Mexico, 1983

1. Silence Is Golden Again

When the Big Guys who run this country decided not to win the Vietnam war and settled into the murderously slow process of diplomatically giving away American lives and prestige, part of this loser effect was the cessation of firearm silencer (suppressor) research and development. For nearly ten years, official or serious work in that area of ordnance was moribund.

All this changed in 1980 with the coming of the Reaganistas' born-again militancy in Washington. It was open season on the doves of social and people programs whose budgets were slaughtered. The hawks roosted high in power, and the trickle-down theory of defense funding became a flood of research and development dollars. Concepts, plans, and new ideas were sought for all sorts of weapons and weapon systems. Silencers and silenced weapons were swept along with this militant new tide.

Accompanying this storm is a new cast of silencer designers and personalities. This current generation of designers realizes that maximum efficiency and effectiveness is a match between weapon, ammunition, and moderating unit. This takes into account all the variables which create that total gunfire sound they wish to suppress or moderate.

The sound of a firearm includes several components:

- the muzzle blast as the hot propellant gases suddenly expand into the atmosphere at the end of the barrel;
- the sound of the bullet traveling downrange away from the firearm;
- the mechanical sound of the firearm's action.

This last is particularly loud with self-loading firearms and is the least noticeable with manual actions. If the bullet travels over the speed of sound, 1100 fps at sea level, there will be a sonic crack. The only was that this can be eliminated is to redesign the barrel as part of the silencer and slow the bullet down to subsonic speeds by porting the barrel. This is not practical with many weapons, particularly with .223 and .30 caliber rifles, because much of the impact power of the bullet is related to its velocity. With some weapons, particularly .22 rimfire and 9mm parabellum using proper suppressor/barrel design or custom loadings, the bullet can be slowed enough to maintain subsonic velocity.

However, the main effect of a silencer is to reduce the muzzle blast, which is the most significant portion of the noise. Muzzle blast is caused by high pressure gases suddenly escaping into the atmosphere as the bullet leaves the end of the barrel. If the pressure can be reduced in either the barrel or a chamber attached to the end of the barrel, there will be less sound generated because gases do not escape suddenly, but are caught and cooled in the chamber. This is basically what a silencer does.

There is a fundamental law of physics which says (Pressure x Volume)/Temperature = A Constant for a given number of gas molecules. In a silencer the pressure is reduced both by increasing the volume for the gases and reducing the gas temperature, i.e., cooling. How well a given silencer works on a given weapon depends on how efficiently these goals are accomplished.

Designers and testers of silencers rate the units in terms of this efficiency in reducing noise levels, among other factors. The sound levels are measured with various instruments, such as a microphone attached to either a voltmeter or an oscilloscope. Both measure the voltage produced by the microphone, and this voltage is proportional to how loud the sound is. Since the voltage is relatively meaningless–it can vary with the efficiency of the microphone–it is compared to a standard and the result is expressed as a ratio.

The next point is that the response of the human ear is not linear. If it were, a sound would seem twice as loud if it generated twice the voltage in the microphone. Instead, the human ear has what is known as a *logarithmic response,* so that a sound that seems twice as loud as another will produce about four times as much voltage on the microphone.

For this reason, sound measurements are compared to a standard, and the ratio given is a logarithmic ratio. The unit used is the *Bel,* or more conveniently, the *deciBel,* which is abbreviated as *dB.* For example, a 3 dB decrease in sound level is one half the original sound pressure level, a 10 dB decrease is 1/10 the original, a 20 dB decrease is 1/100 the original, and a 26 dB decrease is 1/400 of the original sound pressure level. To put this in terms of hearing, quiet conversation is about 56 dB, a handclap about 65 dB, a jackhammer about 120 dB, firing a .22 pistol about 120 dB, and an M16 about 145 dB. Anything over 90 to 100 dB can be rough on the hearing, while levels over 110 dB can be painful.

Sound levels also diminish as the observer goes further from the sound source. That is why a jet engine is hardly noticed a few miles away, but may have a sound level of several hundred dB if it is nearby. The same is true for a firearm. Since the sound level drops according to the inverse square law–meaning *the sound decreases with the square of the distance from the source,* if the sound of a firearm can be reduced significantly, then it cannot be perceived from as great a distance.

Obviously, sound level measurements will vary with how far the test microphone is placed from the firearm. The standard distance of five meters (16.4 ft.) was developed by the Frankfort Arsenal, and unless otherwise specified, most testers use this standard in measuring suppressors. Although the absolute sound levels vary with distance, the ratio or degree of suppression will be constant regardless of the distance. This is why it is more meaningful to talk about the degree or amount of noise suppression than the absolute sound level of the silencer-equipped weapon.

This gives designers, testers, and users a common base for competitive claims as to efficiency in noise reduction. Other comparison factors include alignment, service and maintenance, accuracy effects, size/weight, sighting, dependability, price, practicality, and perhaps most importantly, fieldability, that is, how well it works in the field.

Until the gangster wars of the late 1920s, few people worried about who bought such exotic weapons as silencers or submachine guns. Most Americans regarded them as military hardware, and there was little public interest in these weapons. Then, folks like Al Capone, Mad Dog Coll, Baby Face Nelson, Ma Barker, and others of their ilk littered Middle American main streets with the bullet-riddled bodies of their business associates. Silencers and submachine guns were suddenly viewed as dangerous gangster tools.

It was pioneering refutation of the logic, "Guns don't kill people; people kill people." Crusaders and do-gooders pressured the usual bunch of confused and wishy-washy politicians to outlaw the construction, ownership, possession, or use of such "evil" devices as silencers and machine guns. Illegal possession is a very serious felony under provisions of the 1934 National Firearms Act (NFA), which was amended by the Gun Control Act of 1968 (GCA).

The NFA imposes a tax and registration on the making or transfer of certain types of firearms and destructive devices. This is the major limiting factor on private ownership and use of silencers. Basically, this law requires that a $200 tax must be paid for the transfer or manufacture of a silencer, that any such transaction must have the approval of the feds, and that device must be registered. The penalty for conviction of any violation of, or failure to comply with, any provision of the NFA is a fine of not more than $10,000 or imprisonment for not more than ten years, or both. Naturally, any silencer involved in such violation is subject to seizure and forfeiture.

Under the current law, the following prohibitions pertain to silencers:

- You may not receive or possess a silencer which is not registered to you in the National Firearms Registration and Transfer Record;

- You may not receive or possess a silencer which is not identified by a serial number as required by the Act;
- You may not receive or possess a silencer which has been imported or brought into the United States in violation of section 5844;
- You may not receive or possess a silencer in violation of lawful state or local regulation or ordinance.

As noted in the last point, just because the feds will allow you to have a silencer doesn't mean your local authorities will go along. Restrictive state and territorial laws in twenty-one areas prohibit BATF from approving the sale and transfer of silencers to citizens of those areas. According to a 1982 listing issued by BATF, the prohibited areas include American Samoa, California, Delaware, District of Columbia, Guam, Hawaii, Kansas, Maine, Massachusetts, Michigan, Minnesota, Mississippi, Missouri, Montana, New Jersey, New York, North Dakota, Pennsylvania, Puerto Rico, Rhode Island, and Vermont. It should be noted that in Montana and in the District of Columbia, Class 3 dealers may own silencers, but not private citizens. According to BATF, no dealer in any of the other prohibited areas may own or possess a silencer because of the local and state laws. According to most legal sources, *anything* you put on the end of a gun that "tends to reduce noise" is a silencer and is probably illegal unless you are licensed to own it.

A lot of folks write me and ask how to obtain, buy, or build illegal silencers, and how to obtain, buy, or build legal silencers. My reply is always the same. Forget about illegal silencers. The feds' track record is a superhighway toward busting and hurting the little guy like you because the prosecution's win ratio in these cases is very high. Instead, go directly to *Shotgun News* to check out the ads, find an honest, legitimate Class 3 dealer who specializes in legal transfer sales of NFA weapons to individuals, and let that dealer be your guide.

As of this writing, there are no other laws or regulations. Stick with real law, and don't believe barroom or local police rumors. But laws are only paper. The futility of prohibitive laws is that they are so easily and popularly ignored. Booze was prohibited, so is marijuana. So are all but stringently regulated and registered sales of silencers and fully automatic weapons. Most of these laws are ignored, twisted, or otherwise defeated. I have seen beautifully machined and carefully engineered silencers built equally well by licensed manufacturers, home workshop hobbyists, a former Mob gunsmith, a Special Forces ordnance NCO, and others. In summary, without the necessary government paperwork, permits, licenses, and taxes, the home building, assembling, possessing, or activating of a silencer is patently illegal.

It's quite another thing, though, if you've bought the federal licenses and you're building or buying in large amounts as a defense contractor or corporation. If the sympathy of the U.S. government is with the politics of your customer, you're in even further luck. Yes, folks, the hoary old double standard of the American justice system extends to silencer manufacture and sales. Here's how it works.

Domestically, the men who broker the sales of such exotic military weaponry as silencers refer to themselves as "technicians" or "consultants." Most of them are either headquartered in Washington D.C. or have a contact base there. According to E. Meade Feild, an investigator for the U.S. Customs Service, "We run into them (these semilegal brokers) all the time. They are fully aware of all the laws and the ramifications of these laws, including the political and official climates of government opinion. Some of these people are completely legal; some are completely illegal. Most fall into a gray area. All of them do a brisk business."

The men in the field know the truth of this. Seizures of illegal silencers by the feds have increased by nearly 300 percent in the past two years. Curtis Bartlett, an ATF firearms specialist, is most familiar with the Bureau's enormous collection of confiscated weapons. He says, "I see hundreds and hundreds of illegal silencers that come through here for testing, identification, and evaluation. These include both commercially produced models and homemade units. But the numbers are going through the roof."

Ted Lewis, who used to carry both an FFL and an NFA Class 3 registration under another name, told me, "This entire gun control business is a huge monument to hypocrisy. The USSR and the USA run one-two as the world's largest gun dealers.

"The governments make the rules. They say who builds and who sells the guns. They say what and who is legal and who is not. Want to sell guns to Iran now? Hah! Yet at the time the Shah was deposed, his army had more British tanks than did the armies of the United Kingdom. His air force used America's most sophisticated fighters—the same models that were the latest with the U.S. Air Force at the time," he says with an ironic laugh.

"The biggest arms dealers in the world today aren't guys like Sam Cummings. The real death merchants are the various superpower governments," he says with a snort of bitterness. "You can build and sell suppressors, for example, to anyone with the money . . . and whose foreign policy carbons that of Uncle Sam, of course."

Michael Kokin, an arms broker who runs Sherwood International Import out of California, adds, "The mightiest arms smugglers are the various governments of the world. I'm sorry, it's a cruel world out there. It's the 'haves' versus the 'have nots.'"

In the event you doubt the existence of the weaving spiders of power, stroke your justified paranoia with the spectre of Bohemian Grove, that fantastic summer camp in Northern California, where the Power Control Group communes with nature and each other. For two and a half weeks each year, the very elite of our nation's powerful government and corporate leaders meet there to fine-tune their old boy network and agree on the annual decisions that effect the rest of us.

In 1980, for example, Edward Teller, the unrepentant father of the H-Bomb, told his colleagues at the Grove, "The Soviets mean to take over the Persian Gulf and all the Mideastern oil fields. If there is a small war, a conventional war, we will lose. Unless we have a new beginning soon, I don't know what will happen."

His audience that day included Ronald Reagan, Caspar Weinberger, Justin Dart, William French Smith, William F. Buckley, George Bush, Admiral Thomas Hayward, Willard Butcher, William Casey, and others whose policies reflect the mirror of the Bohemian Grove philosophy, that is, America First at All Costs.

Putting it another way, Tris Coffin, a respected Washington journalist and commentator, has said, "There is a lot more profit in war than there is in peace. It makes a lot of cents to make war, so we do."

Politics and sentiment aside, Coffin is correct and so are the Bohemians—it does make money to make war. And, this simple fact, aided by the rampant Red Menace mentality of the Reaganistas, is why the military weapon coffers are once again open to America's inventors, designers, salesmen, and others with hardware or software to sell.

There is a Big Dollar prize at the end of the fiscal rainbow, and it is still stretching from the White House to the Pentagon. The military hardware makers are tooled up to go to war so we won't have to go to war. One former silencer manufacturer from the Vietnam era told me, "This coming decade will be the most fruitful and lucrative one in the history of ordnance, let alone firearms silencers. We will see more innovative developments and breakthroughs in technology than ever before. I'm excited for the guys who will be working in it now."

Meeting, interviewing, and studying the new generation of silencer men is instructive. For one thing, government technocrats have gotten hold of official silencer design, terminology, and administration. To marketing-minded Hiram Maxim, a silencer was a silencer. The public knew what that meant, even if it wasn't technically accurate. It took the modern marketing genius of Mitch WerBell III and the military's jargon-wound engineers to come up with the term "sound suppressor" to refer more accurately to the function of the silencer. Today, according to my inside sources who attend the R & D briefings about silencer-cum-suppressor technology, the current reference is *low signature weapons.*

Curious about the term, I contacted an old acquaintance at Eglin AFB where specialized small arms arts are still tested and asked him if he had some information about a specific developmental low signature weapon which I mentioned by its operational codename. Here's exactly what he wrote me.

"There isn't much I can tell you, it's a classified project," he said. "Basically, it's a TOP SECRET with modified TOP SECRET. It's an TOP SECRET modified for a slip-in integral silencer, although the barrel is not ported in this version."

Despite all the classification, technology and jargon, when it comes down to working descriptions by the men who use them in the field, these handy accessories are still called silencers. That's what I will continue to call them in this book for the most part. Textually, I will interchange the terms suppressor, moderator, and silencer. So what! The man or woman whose job involves this technology probably calls them silencers anyway. The users care only that they perform well. Academics and lab testing don't impress them one whit. Users live by getting in, getting it done, and getting back out again as quickly and safely as possible.

Despite America's reputation as the world's basement tinkerer and technology genius, our ordnance record has not been outstanding during the past few decades. Tom Kelly, a Vietnam combat vet who was later a Department of Defense civilian employee, says, "Soldiers will throw a weapon away if they don't have faith in it. The individual soldier is the guy making the ultimate decisions on weapons systems, and he's going to throw it away if he can't count on it when the fighting goes down."

Many vets recall other vets trading or losing their M16 in favor of the AK47. One of my buddies preferred the old M1 Garand, another liked the M3 greasegun, another carried an Ithaca Model 37 riot gun rather than the M16.

Sheer magnitude has often rescued the American military machine from its own incompetence. The legendary David Hackworth, a retired U.S. Army colonel, was quoted in this story from Stuart Loory's book *Defeated:*

> I remember a German lieutenant who'd been captured at Salerno. I was guarding him at a POW camp in 1946. He was a real tough-looking kraut and I was a young punk, a pimply-faced kid.
>
> He could speak perfect English, and I was kidding him one time. I said, "Well, if you're so tough and if you're all supermen, how come you're here captured and I'm guarding you?
>
> He said, "It's like this. I was on this hill as a battery commander with six 88 mm antitank guns and the Americans kept sending tanks down the road. We kept on knocking them out. Every time they sent a tank we knocked it out. Finally, we ran out of ammunition, while you Americans didn't run out of tanks."

Interestingly, silencers have never been a part of our ordnance establishment of failure. Like special mission units, exotica like silencers are regarded as a pariah by the military traditionalist who run things. There has never really been a mass-produced, issue silencer, even during the early days of the Maxim models. Silencers have always been an afterthought, a special-order, limited-supply item or the result of the time-honored wisdom of GI field expediency.

The truth is that much of what quiets gunshots in action is not some lab- or machine shop-generated bit of shiny technology. Indeed many of the units are field modifications and expedients of various pieces of ordnance. For example, stories abound of individual and unit "gunsmiths" and self-appointed ordnance experts fashioning silencers from grenade launchers, flash hiders, and other materials. One of the more interesting reports comes from *Devil's Guard,* Robert Elford's fascinating book about Indochina. His narrator, a WWII SS officer who found fun, fame, and fortitude in the postwar Legion, reports:

> I found the muffler-equipped machine guns which we used on so many occasions extremely effective, so long as no prolonged firing was necessary. With mufflers the barrels would quickly overheat. Another shortcoming was that mufflers blotted out the gunsights and tracers had to be used to zero in on the target. After several months of experimenting, Sergeant Krebitz discovered that fairly good silencers could be made from sections of hollow bamboo, padded with wet clay and wrapped in layers of cloth. The result was a clumsy contraption which nevertheless worked.
>
> The soundless death coming from the "nowhere" always shattered the guerrilla morale. The initial shock and the ensuing panic usually prevented the enemy from executing necessary defensive measures. By the time their leaders decided what to do, it was too late for them to do anything but flee or perish. So whenever given a chance we killed in silence.

Any person who has heard, seen, or felt gunfire in combat would know the hair-raising psychological effect of silenced weapons. Elford again quotes his narrator describing his unit's use of silenced weapons with devastating effectiveness in an ambush:

> In such attacks the survivors would disperse and take cover, not knowing where to turn, where to shoot. The sudden realization that the jungle was no longer their ally, that it harbored an invisible adversary who killed in silence, the thought that they might be sitting in the center of a deadly trap, demoralized the enemy. In my opinion all troops engaged in antiguerrilla warfare should be issued rifle silencers. It was the kind of opposition the Viet Minh dreaded: the unknown, the unseen, the unheard death.

Sun Tzu wrote the first real book of military theory circa 400 B.C. In his classic *The Art of War* he writes that of the ways to conquer an enemy, the most desirable is to destroy his mind. That could be a perfect call for the use of silenced weapons. Most modern military experts with combat experience will tell you that in battlefield tactics, maneuver requires strong emphasis on deception, unpredictability, and surprise. That trilogy creates a perfect scenario for the use of silenced weapons.

2. Testing and Evaluation

Unfortunately the criteria and techniques for judging small arms sound suppressors are not standardized nor even validated. The principal, but not the only criterion of silencer quality, is the degree of acoustical attenuation. This requires the definition of some terms and units of measurement.

Muzzle blast is measured in terms of blast overpressure, usually pounds per square inch, or in terms of peak sound pressure level or *peak deciBels.* Although the deciBel scale is nonlinear, we can relate to it when we convert it to *sones,* a unit of loudness. For instance, a 10 dB increase doubles the loudness in sones; a 10 dB decrease cuts the loudness in half, using the same linear units. If a silencer drops the dB level 10 dBs, it is 50 percent efficient. It has cut the loudness in half. If it drops the dB level 20 dBs, it is 75 percent efficient; three quarters of the loudness is quieted. A 30 dB drop means almost 90 percent efficiency. Put a different way, a 20 dB silencer is twice as effective as a 10 dB silencer and half as effective as a 30 dB silencer.

Note, however, that while a 30 dB suppressor is twice as effective as a 20 dB unit, the improvement in efficiency is really only about 14 percent. This is the sort of statistical manipulation that you must watch out for. When in doubt, check the raw dB data, which are usually absolute.

The way in which the dB level is measured, whether the equipment measures peak SPL or not, the microphone type and location, and even the brand of equipment can have a major effect on this numbers game. Other external conditions such as variations between test guns, weather conditions, altitude, and temperature can account for small, but significant discrepancies. However, if the same procedure is followed with the same kind of equipment and environment, then a meaningful comparison between two silencers can be obtained. Uniformity and consistency in testing is vitally important.

What kind of equipment are we talking about? The two basic varieties of silencer testing equipment are the oscilloscope and the sound meter. *Oscilloscopes* are cathode ray tube instruments used to electronically analyze signals. A microphone will convert sound, such as a muzzle blast or a suppressor's acoustical signature, into an electrical signal which will show on the oscilloscope screen as a waveform—a picture of the pressure/time event. This can be measured physically by calibrating the instrument, then reading the screen. This waveform can be photographed from the screen or it can be digitally stored and retrieved for analysis or conversion into hard copy by a chart recorder.

An even more elaborate treatment is obtained by using a digital Fast Fourier Transform real-time high resolution signal analyzer with a digital cassette recorder as a peripheral so that a library of sound signatures can be accumulated and manipulated more extensively by the analyzer.

The advantage to the design lab of these techniques is that much more is learned than just the quantitative peak SPL data. Individual components of the muzzle noise from the suppressor such as blow-by, precursor wave, and projectile noise can be isolated and measured. These can be handled individually in various ways. The mechanical noises

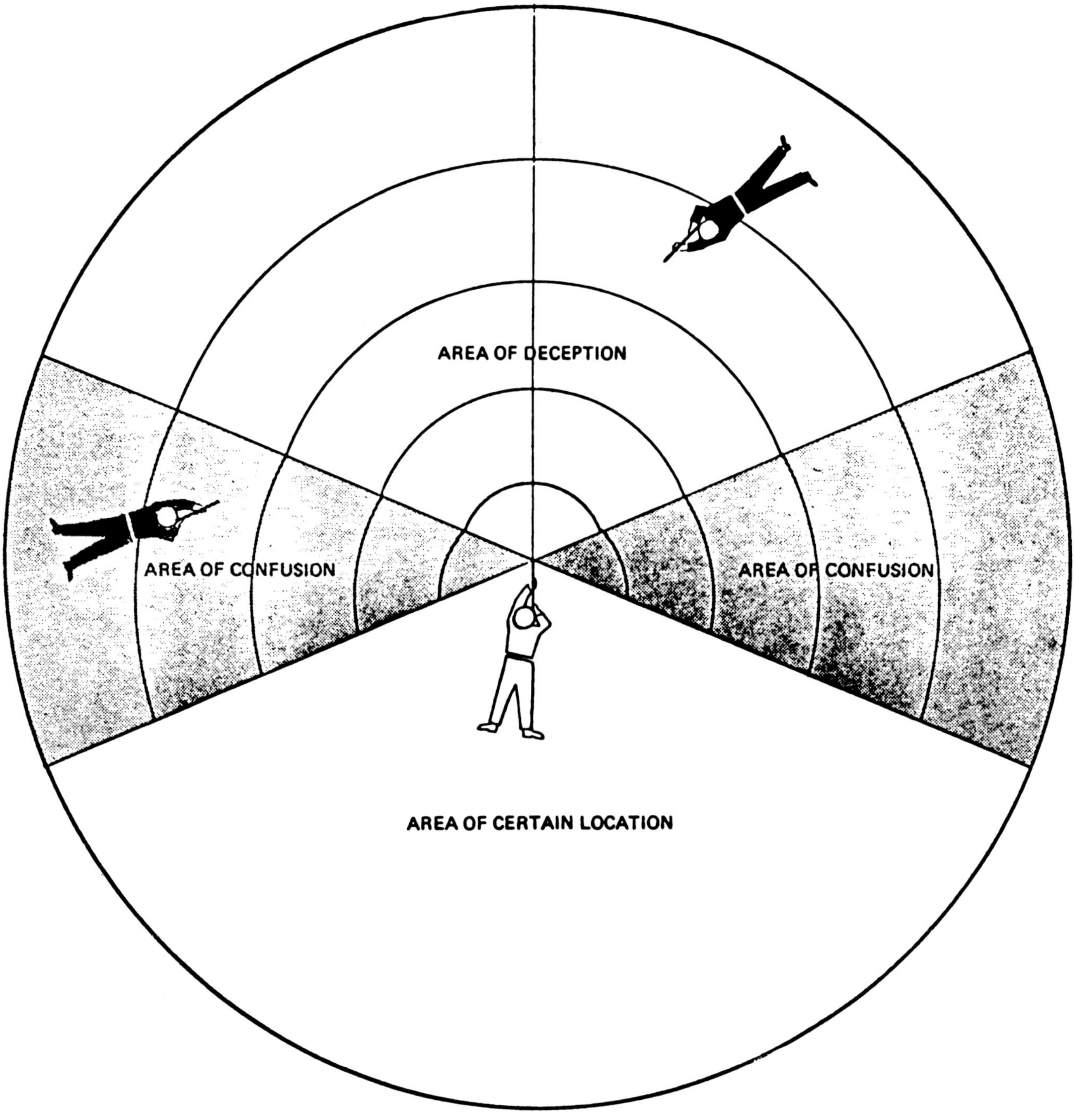

The muzzle blast spread from a suppressed weapon shows deception and confusion resulting from attempts to locate sound origin from weak fixed source at various angles from the source.

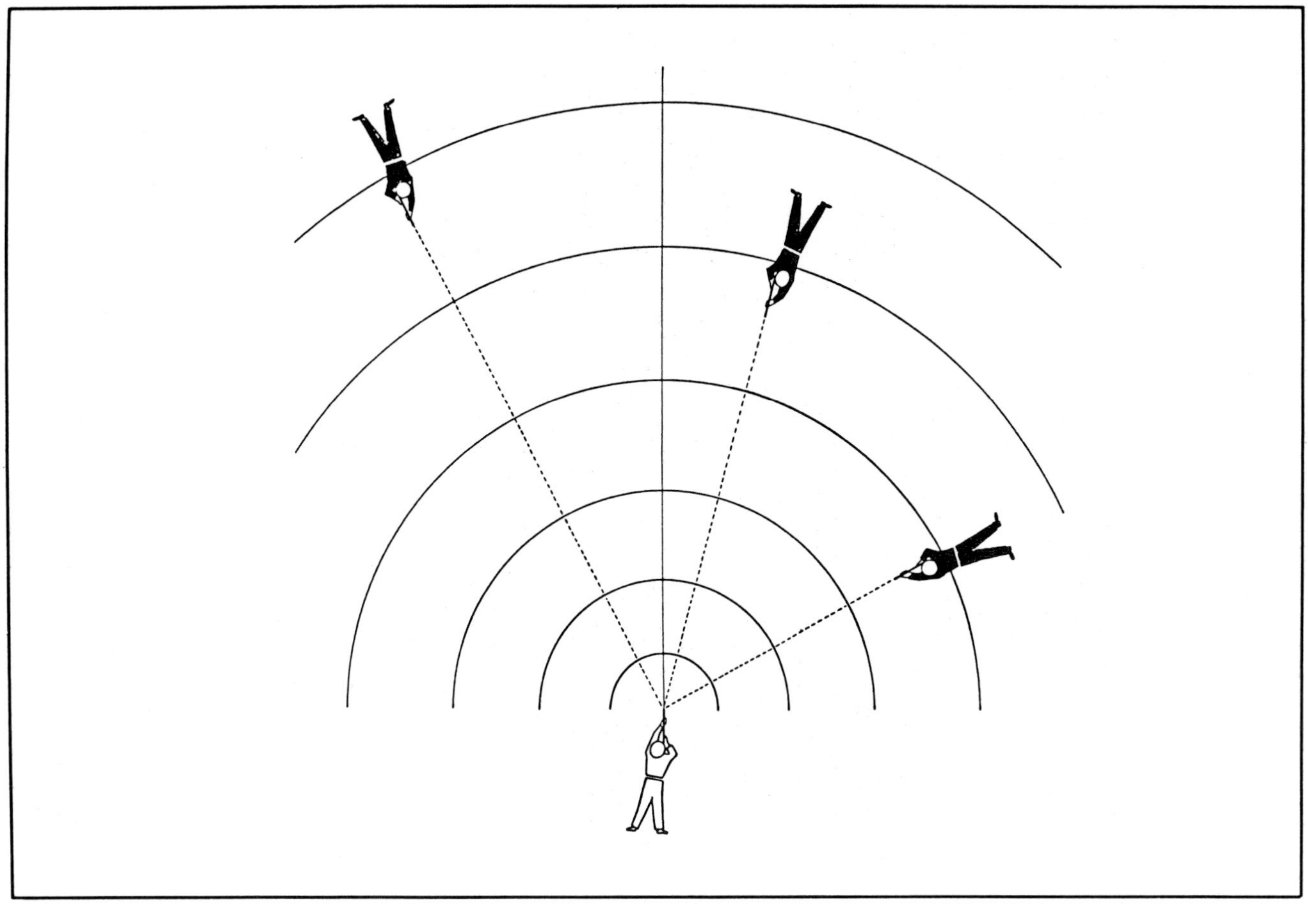

The muzzle blast spread of an unsuppressed weapon shows relatively easy target acquisition of sound from a fixed source. The second source is at right angles to sound wave which locates center or point of origin.

MILITARY ARMAMENT CORPORATION

ROCK ISLAND ARSENAL, U.S. ARMY

Test firing M14 rifle; note muzzle blast wave. This is used as a base comparison for effect of suppressors on the wave.

of the weapon can also be analyzed in terms of their often large contribution to the suppressed weapon's sound signature. All of this analysis can lead to design improvements, especially if computer derived.

For simple quantitative comparison, the instrument of choice is the precision peak SPL meter with a suitable microphone. The meter of choice is the Bruel & Kjaer 2209 and the almost universal choice for microphone is the B & K 4136 quarter-inch pressure mike. This equipment is portable and battery operated. It is also accurate and very expensive. The 4136 mike can accurately measure peak sound at levels up to 168 dB, which is mandatory because 7.62 NATO weapons and short-barreled 5.56mm weapons push hard at that upper limit. A related meter, the 2210, is even more accurate, being a digital Type 0 NBS rated model, but it cannot be used above 160 dB even with the 4136.

General Radio Company manufactures impulse noise meters also and their equipment is used by some military laboratories. At the higher SPLs the GR meters usually show a lower reading than B & K, while at the suppressed levels the readings are quite compatible. Thus, B & K equipment will give a higher dB drop for an actual performance, if the unsuppressed weapon in question is a 5.56mm, 7.62mm, or other heavy caliber weapon. Pistols and submachine guns such as the 9mm usually have unsuppressed peaks in the mid 150s so this is not a problem with them.

If you have read other silencer literature, you might note that some dB levels listed here are higher than those reported in other material. This is because of inconsistency in use of laboratory impulse noise equipment. It seems that a few civilian silencer manufacturers have tested their products with dB meters intended for OSHA type environmental studies, not impulse noise peak SPL measurements. With such equipment, major errors of 30 to 40 dBs are common. Measurements above 140 dB are impossible with such equipment. Even if the meter would read higher, the microphone wouldn't respond accurately.

Let's establish some standards for comparison. Measured at one meter, 90 degrees from the muzzle, a .22 rimfire rifle with normal, high-speed .22 long rifle ammunition has a peak SPL of about 135 dB; subsonic .22 LR, such as Eley Match, has a SPL of 133. A 9mm pistol or submachine gun reads about 155 and a 5.56mm goes 165 from an eighteen to twenty inch barrel, and more from a short barrel such as that of an XM177 or an HK53. Yet a 30 dB suppressor on a 9mm submachine gun gives a reading of 126, while on an M16, using ball ammunition, it reads 137. A high-quality, suppressed .22 pistol is in the 110–120 range. Note that these are all peak readings on the A-weighted scale. When properly written, these would be described as peak SPLs and dB(A).

Now that you are armed with the facts about dB testing, here are the other criteria for judging suppressors. One is muzzle flash. Some silencers kill all flash; some don't. Some suppressors, particularly the baffled variety, will flash intermittently on full auto, whenever the mixture of powder gases and atmospheric oxygen permits. This includes the so-called first round flash. Other suppressors, especially those which involve efficient heat transfer, quench the flame-front of the primary muzzle flash and prevent reignition, so nothing escapes from the muzzle except an occasional stray piece of propellant.

The major point in test and evaluation is to standardize measurements and techniques and use top quality equipment in a uniform fashion. Then the resulting data must be objectively reported in a scientific fashion. Or, as everyone's high school math teacher must have said a few times, "Don't try to compare apples and oranges."

3. The Man Who Brought Us Quiet Killing

H. P. Maxim could have been your kindly next door neighbor, your favorite joke-telling uncle, the nice man who ran the neighborhood store, the president of the local Rotary club, or the chairman of the county fair committee.

Instead, by birth and interest, H. P. Maxim was one of a trio from this famous family name in the gun business. Sir Hiram S. Maxim, H. P.'s father, brought us the first real machine gun, while brother Hudson, H. P.'s uncle, invented smokeless gunpowder, the self-propelled torpedo, and a host of other explosive and ordnance items. Hiram P. Maxim is the father of silencers.

It's an unfortunate bit of history that he is remembered only for his firearm silencers, because this man's memory deserves much more. He was also a pioneer in aviation, ham radio technology, and motion picture photography. A native of Brooklyn where he was born on 2 September 1869, Hiram P. Maxim once said of his hometown, "It's a good place to be from, I guess."

H. P. wrote and had published three books and numerous stories and articles. He also wrote a weekly "Science for the Layman" column for King Features newspaper syndicate for many years. Later, on a dare from his sister, he wrote and sold to Hollywood producers the scenario for the Pearl White adventure film *The Virgin Paradise.* Although an inventive scientist and engineer, H. P. was also a man of letters and a classical scholar.

The world knows H. P., though, for his famous Maxim silencer, the infamous "silent death" design which became the engineering genesis for most silencers since. Among various silencer designs, the Maxim units are of the hot gas type, where hot gases are trapped, swirled, and dissipated slowly through a series of chambers formed inside the silencer's outer tube. This allows those hot, explo-

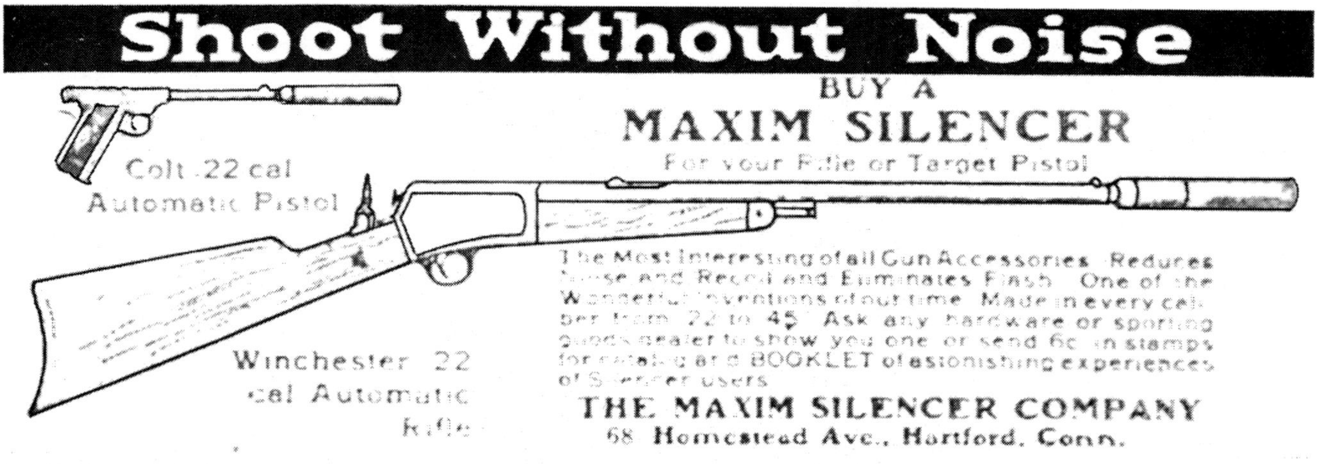

J. DAVID TRUBY

Legal in the good old days, a 1922 *Field and Stream* ad for the Maxim silencer.

Hiram P. Maxim, the man who started the entire silent business.

sive gases to lose their heat slowly, cutting the sound of the muzzle blast as the gases finally get to the outside atmosphere. The efficiency of the holding and expansion chambers is obviously the prime factor in the effectiveness of any particular silencer.

"At first, Father thought the gases had to be whirled around and confined to reduce the noise level," his son, Hiram Hamilton Maxim, a former officer in his father's company, told me. "His first model did just that, keeping the swirling gases inside the silencer chamber. But he soon realized that he needed only to delay those gases to reduce the report."

Although the Maxim Silencer Company was not officially formed until 1908, H. P.'s developmental interest in the device actively began two years earlier. An avid hunter and targeteer, H. P. thoroughly loved shooting, but, always the sensitive soul, he also worried that the noise of his gunfire would annoy neighbors. In 1906, he wrote, "It occurred to me one day that there was no need for the noise. Why not do away with it and shoot quietly!"

For two years he sought a practical way to quiet his firearms. His engineering background gave him the reason for the noise—powder gases and sonic crack. But, how could he resolve that? His memoirs record how he built valves, vents, bypass devices, expansion chambers, and more. None worked. Then, during his customary morning bath one day, his scientist's mind was in a technical daydream watching the water run out of the tub.

Maxim wrote, "I noticed in the bath tub the miniature whirlpool that forms over the drain hole when the plug is pulled and the water starts to run out. There was the familiar little hole down in the center of the whirl, and it started me thinking that here was an exactly similar case to my powder gas and bullet problem. Here was the water in a bath tub, the drain plug being pulled out, and the water was able to run out, but slowly because it was whirling.

"Why should not the powder gases act the same way as the water if they were whirled? The whirling would give them centrifugal action precisely as it did the water and cause a 'hole' to form in the center just as the hole formed in the water. I saw the hole for which I had been looking for nearly two years.

"Immediately, I made a little 'whirling tube' which would catch the powder gases as they burst from the muzzle of one of my rifles and whirl them vigorously. In the center I provided a hole for the bullet to pass through but considerably larger than the bullet so it would not touch. The

MAXIM COLLECTION—CONNECTICUT STATE LIBRARY

A period cartoon showing the military use of the Maxim silencer.

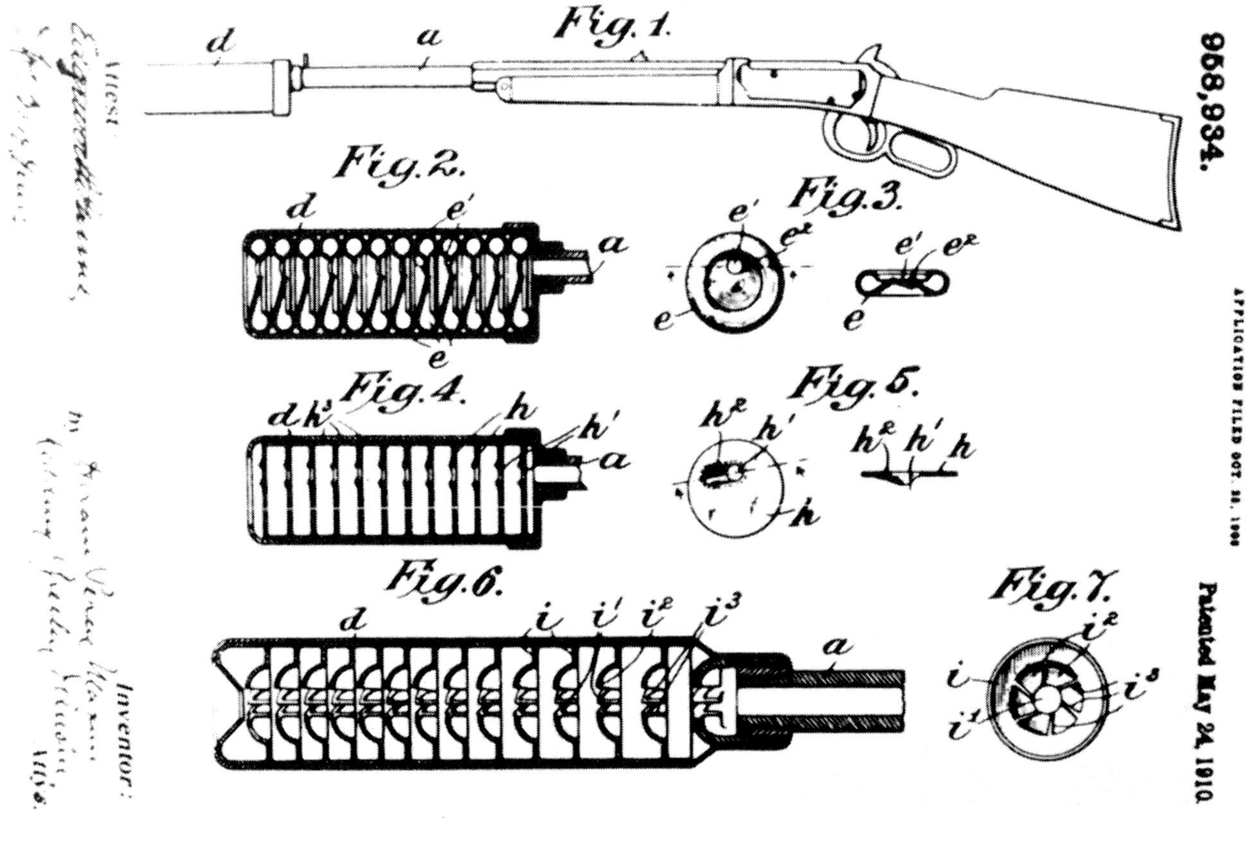

U.S. PATENT OFFICE

The patent illustration for H. P. Maxim's first truly successful silencer.

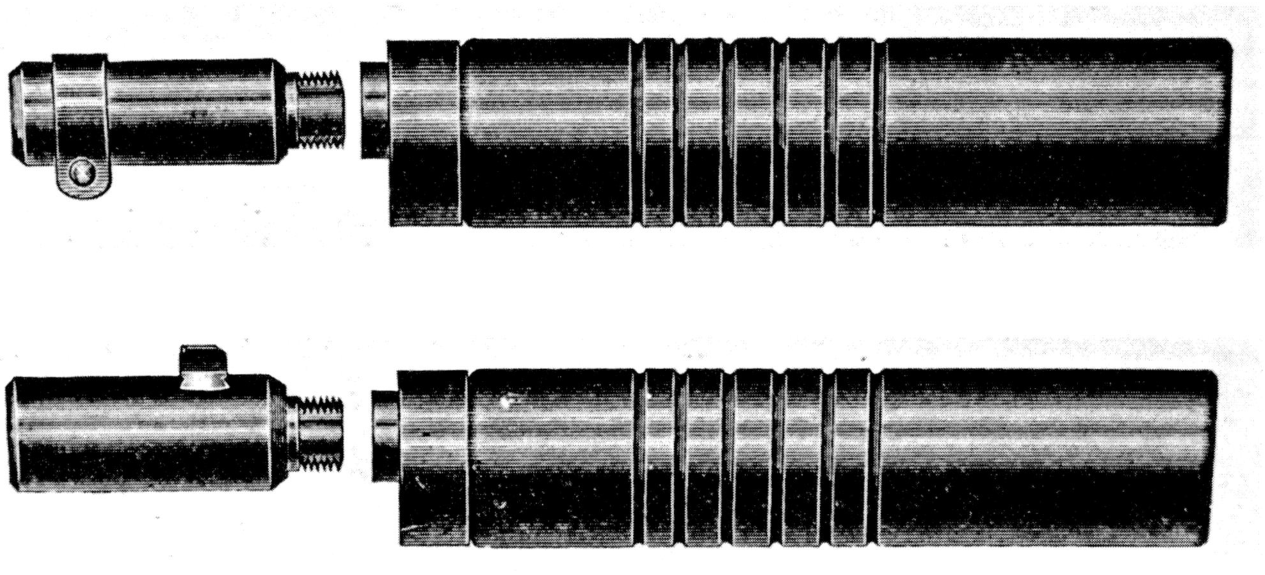

J. DAVID TRUBY

Examples of the clamps and couplers used to attach the original Maxim silencers to the weapons of the day. These couplings sold for $2.50, while the silencers themselves sold for $7.00.

THE NATIONAL GUARD MAGAZINE

Firing the '03 Springfield with the Maxim silencer, 1910. From left to right: Hiram Maxim, Lieut. Col. Richard J. Goodman, and Capt. Earl D. Church.

gases had no escape except through this central hole. Being central they could not possibly get out until they had slowed down. This, of course, meant that they must come out gradually and, in consequence, noiselessly. The first time I shot the rifle I was quite excited. I will never forget the sensation I experienced when I found it was quiet! That was the birth of the Maxim silencer."

His son, Hiram Hamilton Maxim, described that first silencer, built in 1908, as "a little device that looked like a conch shell stuck on the end of his Winchester .30-30. It worked, in that the gases swirled up into that conch, but it didn't work well. My father was a perfectionist, and he knew he could do better. He junked that design and went back to work in the shop."

Maxim's next effort was his entry card to the Ordnance Hall of Fame. His 1909 model was the world's first truly efficient silencer to be designed, produced, and marketed. While it was effective enough at reducing sound, it had two serious drawbacks. It was a sealed unit that could not be cleaned nor have the inner materials replaced. Also, because of the concentric passage for the bullet, the silencer unit obliterated the sight picture on the weapon to which it was attached. Once more, H. P. knew he could do better and did so the following year.

Most experts agree that Maxim's 1910 model was his finest. It was also the most popular sales model and the one which is most commonly encountered now in collections. The major advantage was the off-center passage design which allowed this silencer to be used with the attached firearm's original sights. Another design modification which improved the 1910 model was Maxim's addition of a second physical action. In addition to the circular swirling motion, his new design also forced the gases through sloping holes the length of the silencer tube, cutting down noise even more.

The unit was still assembled from stamped sheet steel and could not be disassembled, though. It was advertised and sold for $3.25 per unit, there being no sales restrictions in those innocent days before the reformers, do-gooders, and politicians had private ownership of silencers handicapped in 1934.

Criminal usage of his silencer was far from H. P.'s mind. He described his successful 1910

J. DAVID TRUBY

The Maxim logo of the 1920s featuring Hiram Maxim's famed Dr. Shush.

H. M. MAXIM

An early industrial ad from the *Hardware Reporter* used art to plug the Maxim silencer for sportsmen's use.

HIRAM H. MAXIM

Hiram P. Maxim holds a Winchester .30-30 rifle equipped with his first silencer. Others are the eight-year old Hiram H. Maxim who later became president of the Maxim Silencer Company; T. W. Goodridge, Maxim's business partner; L. Jenkins, a patent attorney; and Simeon Britt, the machinist who made the silencer.

silencer as being for the man who wanted to shoot targets in the backyard without upsetting the neighbors. Others used Maxim-silenced guns to knock off troublesome garden pests or unwanted alley rats and cats. Innocent America didn't know about silencers used for poaching humans—yet.

W. T. Hornaday, one of the alarmists of his day, called the Maxim invention a "break for every assassin." He also held that union breakers would be able to pop off strikers more easily. There may have been an element of truth in that, as a few Maxim silencers were used to quietly settle labor-management problems during steel strikes in 1910 and 1911. Meanwhile, in New Jersey, two young men used silenced rifles to kill six horses on a farm, then demanded an $800 ransom for their promise not to silently kill others. Police caught them and their Maxim silencers.

These were minor exceptions. For the most part, the civilian users of Maxim silencers were law-abiding sportsmen and target shooters. Indeed, in February of 1909, Hiram P. Maxim presented one of his .30 caliber silencers to President Theodore Roosevelt for use on one of his rifles during the 1909 hunting expedition to Africa. According to the British historian Stephen Critlow, Roosevelt used the silenced rifle on his African trips primarily for camp hunting so as not to disturb other creatures or the quiet of the people in the camp. There is no record of his using it for his record game kills, however.

H. P. had the silencer business to himself despite a volley of unsuccessful would-be competi-

J. DAVID TRUBY

Antigunners in the 1900s. An editorial cartoon (left) in the *New York World* made nasty fun of the Maxim silencer.

THE NATIONAL GUARD MAGAZINE

In 1910, Hiram Maxim (above) demonstrated his marvelous invention for the brass on an '03 Springfield and made the cover of the *Guard* magazine.

NEW YORK PRESS

A political cartoon (left) of 1909 featured Teddy Roosevelt hunting political foes using a rifle with a Maxim silencer—easy publicity.

tors. One of his earliest competitors was his father, Sir Hiram S. Maxim, who lived most of his adult life in England. Sir Hiram was granted a British patent for a device to silence a machine gun's operation just two days prior to his son's own U.S. application for his silencer design. According to the late Major F. W. A. Hobart, the noted British firearms historian, neither design went beyond the plans and prototype stage. The son's later designs obviously did, however.

H. P. became Mr. Silencer, or when he satirized himself in his company's sales brochures and ads, he became Dr. Shhhhsh. The man was an all-pro(motional) genius.

When he was fifteen, Maxim entered the School of Mechanical Arts at the Massachusetts Institute of Technology. He was graduated in 1886 when he was seventeen, the youngest student in his class. He went to work for the Jenny Electric Company in Fort Wayne, Indiana and later for the Sun Electric Company of Woburn, Massachussetts, a subsidiary of the Thompson Electric Company, one of the firms which eventually formed today's General Electric Company. In 1892, he became superintendent of the American Projectile Company in Lynn, where he was increasingly interested in the development of the gasoline engine.

By 1895, he was in Hartford developing the Columbia car, as chief engineer of the Electric Vehicle Company, organized by the Pope Company. In 1900, he left Hartford and went to Pittsburgh as vehicle motor engineer for the Westinghouse Electric and Manufacturing Company. In 1903, Maxim returned to his old post as chief engineer of the Electric Vehicle Company.

By 1906, Maxim realized that the Electric Vehicle Company was going to go out of business. With a friend, he formed a partnership to build the Maxim-Goodridge Electric car. They were unable to find either serious financial backing or potential customers so the partnership was disbanded after one prototype. It was then that Hiram Maxim turned his attention to the firearm silencer. In 1908 his first official business name was The Maxim Silent Firearms Company.

Despite his original desire to produce a silencing unit to make the noise of plinking less bothersome, the nasty reality of business quickly and loudly shattered his dream. Although hunters, sportsmen, considerate plinkers, and even a few poachers bought Maxim silencers from their local hardware and gun shops, there were simply too few

J. DAVID TRUBY

Hiram Maxim, son of the famed inventor, with one of his father's tubes—in pristine condition—on a favored .22 rifle.

units being sold to the American public to turn a profit.

"My father was finally convinced that the military market was his only mass market," H. H. Maxim told me during a visit to his home some years ago. "He outlined in a series of handwritten notes the potential military uses including sharpshooting (sniping), guard harassment, destruction of supplies, and, interestingly, marksmanship training.

"A lot of soldiers never learned to shoot well because they feared the awful sound of the explosion of that .30/06 round in the Springfield, and, later, the M1. My father wanted to put Maxim silencers on the weapons used by those nervous trainees. It was a good idea."

Ironically, that same concept—silencer-equipped training rifles—was part of the concept of the U.S. military in its Joint Services Small Arms Project (JSSAP) in 1984.

As a result of Maxim's military marketing thoughts, his 1912 model silencer was designed with soldiering in mind. Several of these units were tested by the Army with their '03 Springfield. The silencer quieted the muzzle blast, but did nothing to stop the sonic crack of the bullet down range. Tests run during the late 1960s at Philadelphia's Frankford Arsenal used the Maxim Model 15 silencer, similar to the 1912 unit, with an M1903 Springfield rifle. These tests used subsonic ammunition which eliminated the ballistic crack. The test report noted that the Springfield rifle and its Maxim silencer was "one of the better units tested at Frankford Arsenal."

But back in 1912, despite appreciation for the Maxim silencer's effectiveness, the U.S. Army wasn't buying in standard issue amounts. It was peacetime with slashed budgets and a tight economy. There was no powerful defense lobby and no Pentagon hawks to buzz around the Hill squawking their war cries of "Stronger defense; more weapons!" H. P. Maxim was on his own.

As usual, special mission items like silencers were lumped under "miscellaneous," which meant that H. P. sold few silencers to the military. Some of those few 1912 models went with Gen. John Pershing's expedition into Mexico to chase after Pancho Villa. Sixteen of Pershing's sharpshooters carried Springfields equipped with the Maxim silencers. There is no record that these units were used successfully, but then, there is little to document that any of the Pershing mission was more than a symbolic success either.

From Gatling through Maxim through Lewis through Thompson, Johnson, Stoner, Ingram, et al., the U.S. Army has been amazingly closed-minded about its weapons systems development. Unable to sell to the home government, American inventors go elsewhere. Hiram P. Maxim was no exception. His sales agent, Joseph Keegan, easily sold Maxim silencers all over the world, with shipments made to China, Japan, Mexico, England, France, Belgium, Russia, and South America.

The most obvious sales point was the lack of sound. H. P.'s favorite demonstration involved mounting a company business card in a special holder set several inches from the muzzle of an unsilenced gun. When the gun was fired, the muzzle blast blew the card to shreds. Then, with a dramatic flourish, H. P. would mount his silencer to the rifle and repeat the demonstration. This time there would be a mild click and a neat hole in the center of the calling card.

"No noise, no fuss, no muzzle blast," he would tell the startled audiences who had to be shown the card for proof the gun was actually fired.

By the time World War I primed to the shooting point, the Maxim Silencer Company was producing a line of quality silencers in calibers ranging from .22 through those large enough for machine guns. Despite the popularity of the Maxim silencers, their deployment during World War I was not widespread. Sniping and trench harassment constituted most of the use; a favorite trick was sharpshooting officers, at which both the Germans and British excelled.

When World War I really opened in Europe and it became apparent that America would be involved despite the politically neutral stance taken by President Woodrow Wilson, the Maxim Silencer Company's business success took off like a rocket. They ran production overtime for the military markets eagerly gobbling up their firearm silencers, gas grenade casings, automotive mufflers, and parts for bayonet scabbards. They sold at home and abroad; money was made.

By 1917, selected sharpshooters of the American Army were armed with Springfields equipped with the Maxim silencer. Some of these men and their weapons made it to Europe for our glorious 180 days of action in the Great War. But, there is little in the official record of American silencer success in that action. The enemy was impressed, though, as the Germans dubbed the British silencer-equipped Enfield rifles, using Maxim units, "The Whispering Death." Meanwhile, for the American GI, his Maxim silencer-equipped Springfields survived their post World War I cosmoline nap to serve snipers again in World War II and Korea.

But, in 1918, after the war had ended, the pany had financially quiet times until 1922 when the industrial silencer market took hold. H. P.'s firearms designs were modified to provide mufflers and silencers for huge industrial machinery. By 1925, the firearm silencer branch was discontinued as being unprofitable. Two years later, the industrial silencer business had the organization back to its wartime success level.

In early February of 1936, H. P., then sixty-six years of age, left New England with his wife for a West Coast holiday. He became ill on the train and a very sick Hiram Maxim was examined at La Junta, Colorado for what has been described as "a fatal infection." He died in Colorado on 17 February 1936.

4. Hear No Evil

Firing an M16 in the old-line residential Garden District of New Orleans on a sunny, warm, spring afternoon is not listed in the Chamber of Commerce tour guide. But then, Don Walsh didn't write that guide.

One evening, residents in the famed Hotel Maison de Ville in the city's French Quarter didn't share the professional delight of the men firing a Beretta model 70 pistol in one of the elegant suites. But then, Don Walsh was there.

Don Walsh invented and now produces silencers so effectively quieting that we easily "field tested" all sorts of weapons in his city without arousing any of the citizenry—nor the police, the militia, or anyone else for that matter.

A thirtyish bachelor, Walsh is the prime mover and shaker for Interrand Corporation, a suburban Washington company with the potential to become the prime supplier of firearms silencers to the world's military and police. According to many of his peers, this man's innovative, yet simple and efficient, silencer designs place him in line to succeed the legendary Mitch WerBell III as *the* name in silencer state-of-the-art.

Walsh acknowledges the comparison, saying, "Everyone in the clandestine weapons field owes a huge debt to Mitchell WerBell III. In addition to the technical expertise and innovative designs he and Gordon Ingram brought to this field, he contributed something more.

J. DAVID TRUBY

Walsh tests one of his Larand units on a sniper rifle.

J. DAVID TRUBY

J. DAVID TRUBY

In the indoor test facility, Walsh checks an M3A1 sporting a two-stage Larand suppressor (two views).

"More than anyone, he elevated awareness of the legitimacy of silenced weapons in the small arms arsenal. Historically, these special mission weapons have been a last-minute, modified afterthought. WerBell literally brought practical silencer design and use out of the closet. He made the genre legitimate. He's why the rest of us are here today.

"From a technical standpoint, WerBell's patents speak for themselves. There are technical benchmarks in the history of practical silencer design. Maxim started it, WerBell refined it, and someday I hope some young new technician might say that Walsh perfected it," he adds with a hearty chuckle.

Walsh approaches silencer design pragmatically; he is not at all interested in gimmicks or James Bondian toys. "My designs and finished products are designed to be used in the field under all conditions," he says. "My criteria for a good silencer are simple: simplicity, strength, low or no maintenance, indefinite field life, no reduction in weapon accuracy or bullet velocity, plus the greatest possible reduction in sound. In addition, I work to make the size and weight of my silencers as minimal as possible commensurate with the specific weapon in question."

So far, Walsh has succeeded in all areas according to all tests to date. I have examined, tested, and used his various silencer/weapon combinations.

They're good, no question about that. From my empirical observations I would say that the Walsh silencers generally perform about 30 to 40 percent quieter than the Military Armament Corporation silencers we used for comparison. Weapons tested included the M16A1, Beretta pistols, various Heckler and Koch pistols and assault rifles, the M3A1, Ingram M10 and M11, plus a Thompson Centerfire in massive .45/70. No ear protection was needed; the man's silencers are good.

In addition, Walsh's silencers have been examined and tested by various weapons and military experts. Their evaluations have been as enthusiastic as mine. Don Walsh has a new idea—in the world of silencers he has designed a quieter mousetrap. He didn't do it by accident, either. He's an educated, experienced, practical man with a delightful sense of humor.

Walsh attended the University of New Orleans at various times between 1968 and 1980, studying political science, then working as a research chemist. He is a native of New Orleans, which he chucklingly refers to as a city of one-way streets, two-way men, and three-way women. This inventive humorist says that his love for meerschaum pipes, Upmann cigars, great food, and rare liquors even comes ahead of his love for fine firearms and their (his) silencers.

Yet Walsh brought to suppressor design some very definite ideas about the proper design of the silencers men need in the field. Unlike the "see-a-problem-solve-it-from-the-field" type of designer, Walsh knew that homespun, cracker-barrel ingenuity wasn't the total answer. Instead, he looked at the science and technology theories involved in

INTERRAND CORP.

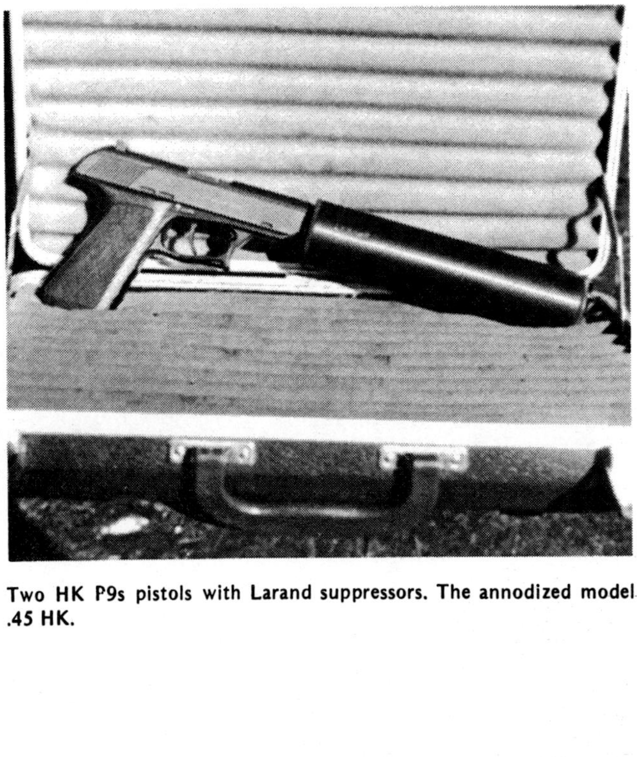

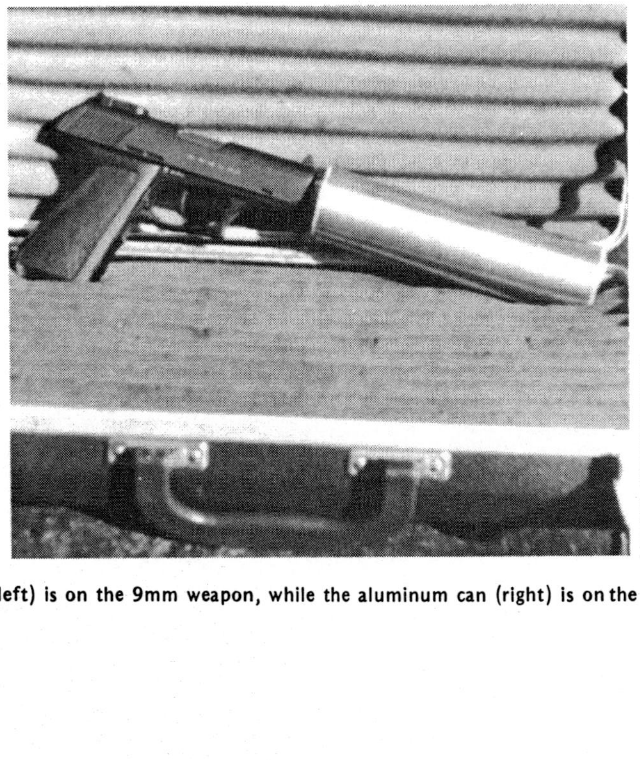

INTERRAND CORP.

Two HK P9s pistols with Larand suppressors. The annodized model (left) is on the 9mm weapon, while the aluminum can (right) is on the .45 HK.

problem solving before cranking up a lathe. Walsh has obviously done his homework. He put his scientist's mind to work, studying to become knowledgeable in ballistic and acoustic engineering, metallurgy, structural design, and even the nuts and bolts of machining.

"There was just too much technology at stake for me to treat this like some basement tinkerer's hobby. I studied the appropriate sciences and engineering, then began to survey the literature," he says. "As both a past user of silencers in the field and from my research, I concluded that for a given volume silencer, a wire mesh type of packed silencer is inherently superior to a baffle type. This was the same conclusion AMF came to in their very valuable research and I concur, now that I have thoroughly tested variations of both types of designs."

Walsh adds, "The mesh system is far superior because of its dynamic interaction with the impulse noise of a firearm. Hot gases and burning propellant are expanded and cooled while acoustical energy is lost by mechanical transfer to the mesh. On the other hand, the baffle design is static, with no interaction.

"Baffles and chambers are inferior because they serve only to delay the escape of these sound-causing gases, which are cooled then by adiabatic expansion."

He says that in addition to being a far superior silencing medium, his mesh design also allows a much smaller package to perform better than a larger baffle design. His proven concept now allows the use of a smaller, lighter silencer for larger caliber weapons than previously encountered—a definite advantage in the field.

"I was also able to overcome the only major disadvantage of a mesh silencer. The old method was to stack inside of a tube washers cut from woven wire cloth of a suitable wire diameter and mesh size. The inside diameter of the washer stack formed the bullet path and was typically .025 to .050 inch larger than the projectile diameter," he says.

"However, vibration and blast allow individual strands of loose wire to work loose and into that bullet path. When you have 240 washers, as in the typical unit, the possibility of wires getting into the bullet's path are quite certain. If enough wire debris is in that path it will throw off the accuracy of the projectile. This is why, for example, the High Standard HD used by the OSS and later by the CIA, as well as the M3A1 silenced by Bell Labs, have to be repacked so often.

"That situation defined part of my problem—to find a packing with the acoustical and thermal/mechanical properties of a wire screen washer stack, but without the loose ends. I finally located a prime manufacturer of metal and plastic textiles who was willing to work with me on the metallurgical engineering. Soon, samples of the type of packing I wanted were fabricated for field trials. The results were so positive that I immediately ordered tooling and dies for my custom mesh pieces that would be the guts of my design."

His original custom design—seamless, knitted copper wire mesh—was formed into what Walsh calls "donuts" in the proper size for each caliber. The mesh donuts and various spacers are simply placed in one of Walsh's specially designed tubes, end caps are screwed in place, and the unit screwed on the end of the weapon in question. The result is one of the quietest weapons in the field.

For production purposes, his original system has been superceded by a simplified refinement which Walsh calls "GI-proof," meaning it is now field ready for the troops. He adds, "I am now using an exotically machined central core passage in all my units except the .22 rimfire models. This change allows absolute simplicity in manufacture, plus control of alignment at a much lower cost than before."

The professional firearms design field is not a large group of people, and relations between the principals are not always as pleasant as someone like Dale Carnegie might wish. Indeed, there are some downright hostile feuds between contemporaries in this field. The silencer fraternity is even smaller, and following the cheerful lead of pledgemaster Mitch WerBell's reputation, Don Walsh agrees that his field has none of the bitchiness that swirls about in the other areas of less than corporate ordnance. In fact, he speaks very highly of his competition. Men like WerBell, Dr. Philip Dater, Jonathan Arthur Ciener, and Reed Knight all get his highest professional approval and esteem. He adds, "I may disagree with them on a few technical details, but these men are all top professionals and I respect their work."

Walsh does have strong feelings about silencer technology, and isn't shy about stating them. "I am very averse to porting barrels. Usually, porting, venting, wipes, and the like are merely

DON WALSH

Interrand suppressors in attache case headed for military trials.

INTERRAND

The Larand suppressors, 9mm HK P9s laying down with the .45 model leaning on the case.

J. DAVID TRUBY

The action of the suppressed Thompson. Note the ART mounted on the weapon which Walsh calls "a real beast to shoot, but quieter for it."

J. DAVID TRUBY

Master suppressor builder Don Walsh with one of his Larand units on a Walther MPK 9mm smg.

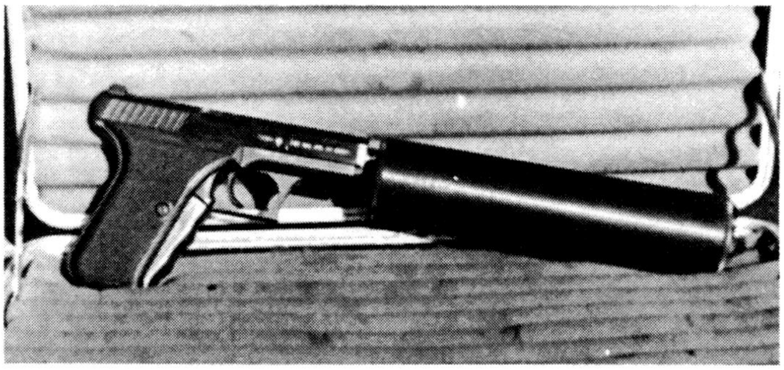

INTERRAND CORP.

The 9mm HK P7 with a Larand suppressor.

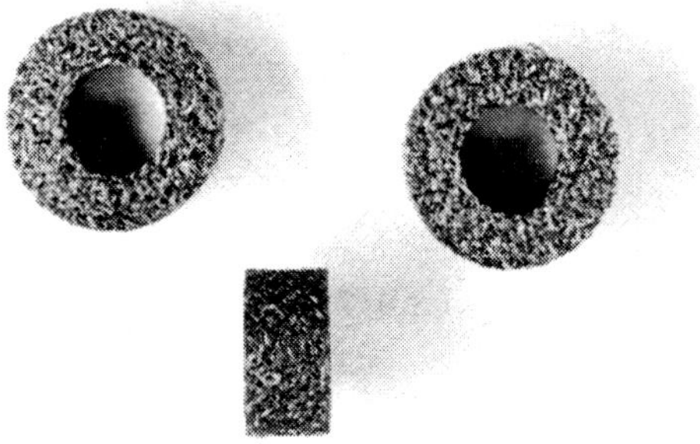

J. DAVID TRUBY

Wire mesh donuts for the Don Walsh suppressors.

J. DAVID TRUBY

Quiet fun on full auto in a residential part of New Orleans. Walsh blasts away, but quietly so, with his suppressed Sterling.

J. DAVID TRUBY

Camera-shy operative tests one of Don Walsh's prototype test suppressors on an HK pistol.

crutches needed to bolster a basically ineffective system so the sound signature of the weapon is at least reasonable," he explains.

"If a silencer is designed and produced properly in the first place, there is absolutely no need for porting or wipes. About the only time you would even consider porting is when you are using a weapon firing the 9mm ball round and you can't get subsonic loads."

There are few weapons Don Walsh says he cannot effectively quiet, moderate, or distort, even large caliber centerfires. But two domestic firearms really draw his ire when he gets requests to turn out a silencer unit for them. He says even some police units want these weapons silenced.

"The AR7 and the American 180 are bad designs to begin with, and I refuse to stock them myself. If I couldn't talk a customer out of it, I guess I could put a can on the end for them, but I wouldn't be happy about it," he adds.

Walsh also has tough words for the various silencer kits being advertised through *Shotgun News.* He says that despite the pious wording about federal approval needed for assembly, these kits mostly end up in an illegal configuration. Most of those he has tested are also highly ineffective.

"I'd advise anyone to stay clear of most of the silencer kits. Most are expensive trouble for you and your wepaon. Not only are the majority poorly designed and manufactured, but shortly after the mailman or the UPS driver brings your kit, you're liable to have friendly ATF boys at your door. If you want a silencer, pay the money to do it right and do it legally."

Asked the difference between a silencer and a sound suppressor, Walsh replies, "Mostly it's semantics, something important only to the arcane technical minds who have nothing to do but worry about things like that.

"Technically, no silencer really silences; it only quiets or suppresses the noise of a gunshot. Hiram Maxim, the man who started all of this by inventing the first practical silencer, called his device a silencer, and that's good enough for me."

Walsh did not get into silencer work by his own design, nor did he slide in by the usual military contact door. He has one degree in political science, another in synthetic organic chemistry, plus he wanted to be a science-fiction writer, so he wrote, sold, and had published his science fiction stories. He has also published as a chemist in the professional journals of that field.

"My entry into the silencer field was a business progression," he notes. "I gravitated from collecting to selling sporting weapons as a hobby, to starting a business, to handling surplus military weapons then conventional military small arms, then weapons covered by the National Firearms Act."

He freely admits he is in the design and manufacture of silencers as a profit-turning business, adding, "I had a Class 3 license in 1975 but soon found that manufacturing is the only way for me to make a real business in this field. I recognized that the demise of MAC (Military Armament Corporation), which started to fall apart when Mitch left the company in 1972, created a real void in this field of military design.

"Some of my earliest work was converting the old OSS/CIA stuff over to my system and seeing how much more improved it was. I refitted an HD, an M3A1, the SOE's Welrod, and some of the modern weapons for early customers, who were usually friends with Class 3 licenses."

He began doing more custom silencer work for both collectors and Class 3 dealers, producing units for their weapons or designing and producing integrated units by matching weapon and silencer.

"It was fun," he adds. "But there was no real money in it, and I like the things in life that you must have money to enjoy. Since someone always seemed to be at war with someone else, I turned to police and military markets in 1981."

There are three markets for silencers: individuals, domestic government, and export sales. The first includes collectors and other private citizens who can qualify for and afford the $200 per unit federal transfer tax on top of the purchase price. Nobody ever got rich with this, and no serious silencer manufacturer considers it a prime market. The government and export markets are something else, though, especially with the Reaganista defense spending philosophy dominating the western world.

Mallory Engler, a former military officer and counterinsurgency warfare expert from England, says, "Vietnam and some of the paramilitary and antiterrorist activities since May have finally convinced the traditional military mind that silencers have a legitimate and most useful special mission role in a variety of operations. This acceptance really opens a broad market for silencer makers."

Engler adds, "The important thing is to demonstrate the devices, as many of the government officials who make the purchasing decisions are men who have never been in the field or actually seen or heard a silencer in action. They must be shown."

Don Walsh agrees with this wholeheartedly, saying, "Domestic marketing is mostly a matter of getting my silencers in the hands of the decision-makers who must see and 'hear' my technology and capability. Hearing this sound of relative silence, if I may be a bit literary, is usually all it takes. It's not ego, but fact; my silencers sell themselves because they meet the need.

"I show my devices to police and military officials and demonstrate their efficacy in terms of command and control situations. That's about all it takes. There are so many ways silenced weapons may be used to save time, money, hassle, and even lives."

The official U.S. attitude was notoriously anti-silencer during the Carter administration. Despite the obvious advantages of silenced weapons in anti-terrorist, counterinsurgency, military/police, and training operations, our government was officially opposed to the devices and concepts. Not only was this regressive policy injuring domestic research and development, the State Department hindered export sales.

"This was the era in which I was conducting most of my own R & D," Walsh says. "It was a good time for me to experiment. Now, with the Reagan administration's emphasis on fighting back against global terrorist activity, the emphasis is on the proper tools for such combat."

Walsh notes, "If my business has something to gain from this shift in policy, so does the ultimate freedom of law-abiding people everywhere. That's especially true for those who now live in fear of the terrorist who is able to move and strike with impunity, unconcerned about inadequately armed police and security forces with their unsophisticated equipment.

"Dozens of 'field laboratories' from Africa to South and Central America to Southeast Asia have taught us that silencer-equipped weapons are the finest antiterrorist weapons around."

Although he is a corporate member in good standing of the American Defense Preparedness Association, the Association of the U.S. Army, and other military/industrial organizations, Walsh admits his name or that of Interrand is not yet in the defense household word category as are Winchester, Colt, and Smith & Wesson. But in the world of silencer design, he is already on his way to a good season and sure stardom.

As Walsh says, "Interrand Corporation is the only full-time military small arms sound suppressor manufacturer in the United States today. We want to develop consistently superior sound and flash suppressors and to promote broader application of this technology.

"Our silencers are designed to meet stringent military requirements for a whole new generation of silenced weapons. Ours are smaller, lighter, simpler, stronger, cheaper and quieter. We accomplish all this without requiring maintenance or replacement of components and without affecting accuracy or velocity. Ours are the benchmark against which all other silencers must be judged."

He grins, then says, "In this business, silence goes a long way, so I'll just call myself the uniquely quiet American.

5. The Baffling Radiologist

I couldn't do it.

Despite the urging of the toothless old gunsel, who seemed ancient enough to personally know the latter, I would not compare Dr. Philip Dater with Dr. Richard Gatling, another famed medical man turned ordnance designer.

Let's leave those comparisons and alliterative titles to the hack writers.

Instead, you may remember Dr. Philip Dater from my book *Quiet Killers II* (Paladin Press). He's the man pictured there with the timber wolf crossbreed, and he's also the man who designs some of the finest suppressors around today. He doesn't work for the CIA, FBI, DIA, or any government organization. He designs and builds suppressors mostly for ordinary people who legally want or need a weapon that doesn't make much noise.

Unlike other suppressorists, Phil Dater does not actively seek government contracts, saying, "I've probably sold more than a few to government operatives. Mostly, though, my customers are ranchers and farmers who want to take out pests without disturbing the livestock, the family, or our native tranquility.

"I do work for target shooters who want backyard practice without making the neighbors think the Russians have landed. Also, I have customers who are collectors of exotica and militaria seeking examples of suppressed weaponry."

Dater operates under the licensed name Automatic Weapons Company in New Mexico, a legitimate, licensed Class II manufacturing operation. His work and his products are universally praised among the suppressor community, designers and users alike. Phil Dater is highly regarded for his ideas, workmanship, and products.

"I work with serious buyers who are willing to pay for a quality product. I am dedicated to continued quality and service for the owner. My units are designed and manufactured so the owner can maintain them himself. In fact, I include all the special tools needed, along with a detailed instruction manual with each unit. I'm very much an owner/user-oriented businessman."

PHIL DATER

Phil Dater has standardized his 9mm suppressor as the SG9, adaptable to any 9mm submachine gun. It's shown here on a S & W M76.

PHIL DATER

Phil Dater's highly respected RST sound suppressor on a Ruger RST-4 pistol. The Dater modification uses adjustable sights.

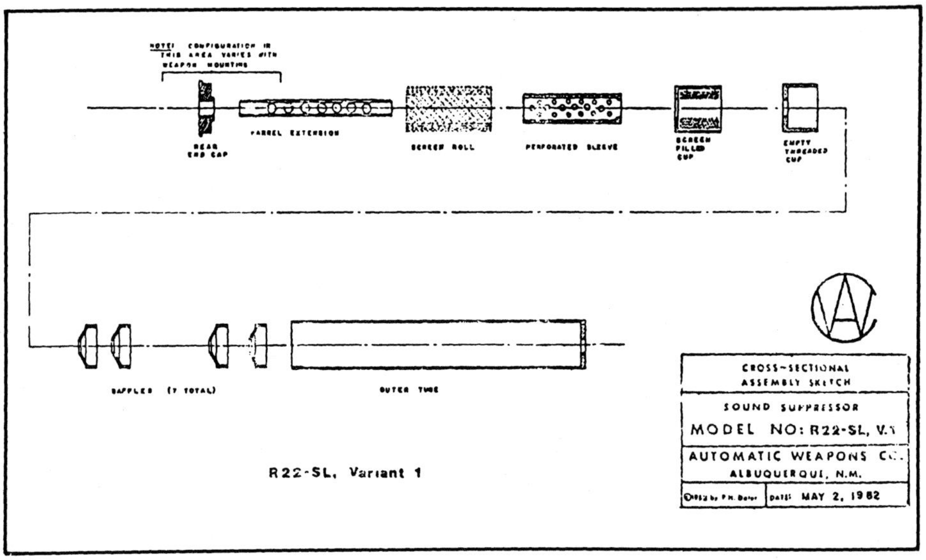

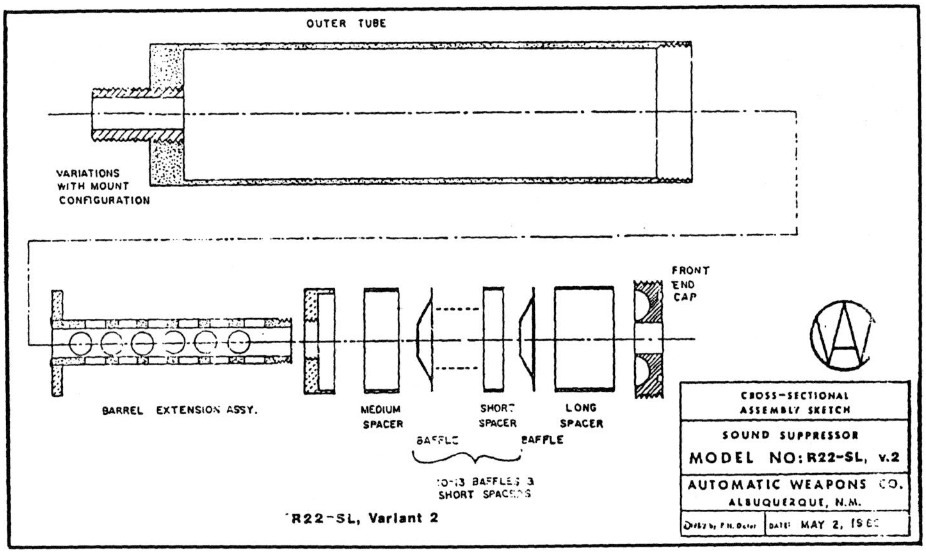

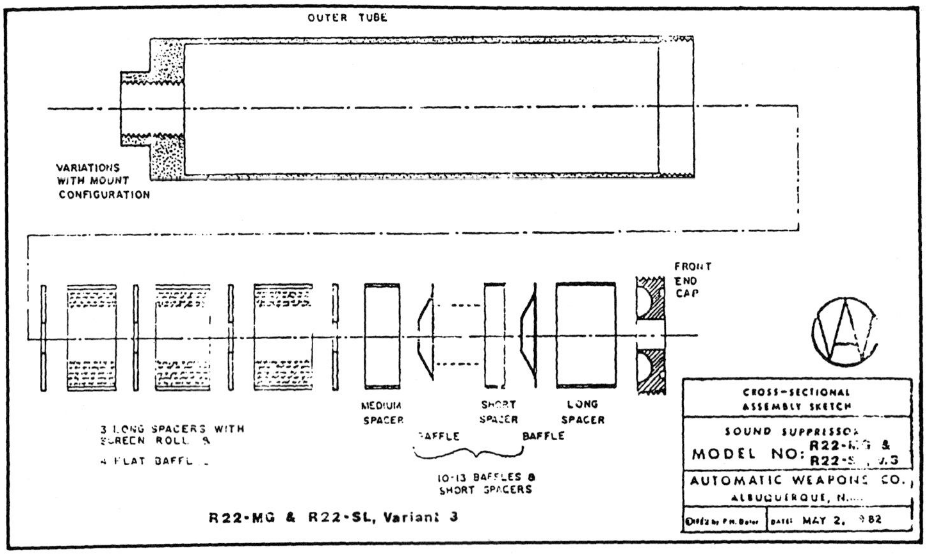

PHIL DATER

Three Dater suppressors: R22 (top) and R22-SL (center) for externally threaded barrels; R22-SL for internal threading.

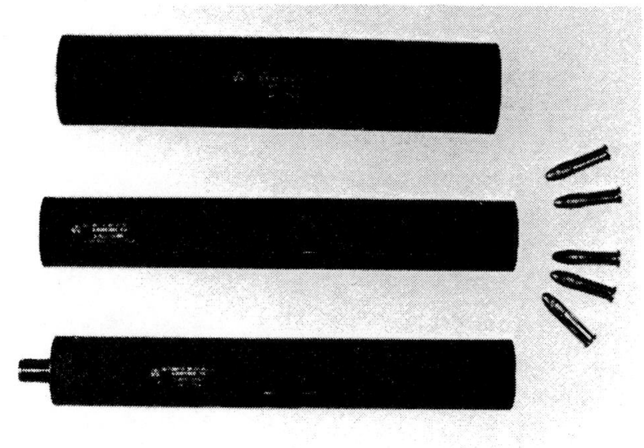
PHIL DATER

Three variations of the R22-SL suppressor.

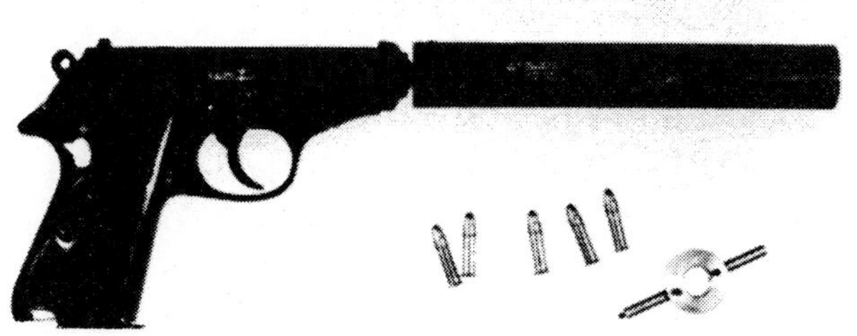
PHIL DATER

Ine Dater "James Bond Special," a model R22-SL internally threaded into a .22 Walther PPK-S.

Quality is his watchword, as Dater uses only top quality ordnance steel and aircraft-grade aluminum alloys for the manufacture of his units. All of the steel parts, including the interior ones, are blued with a deep black oxide.

"However, the steel baffles for the .223 and .30 caliber units are usually stainless steel because of wear and corrosion resistance. The external aluminum parts are black anodized for both aesthetic and protective reasons, while I leave the internal aluminum in its natural state for thermodynamic reasons," he says.

Dater is unique in that he offers a very straightforward warranty for a year. To his knowledge, no other major manufacturer does that without charge. It's worth it, though. Phil Dater's units are not inexpensive. Rather, like their manufacturer, they are very classy.

As you may already know, Phil Dater is a physician, specifically a radiologist. In the summer of 1980, he turned away from the corporate world of a large, multi-specialty group in Albuquerque to work in the country.

"So far I have resisted all efforts and recruiting to get me back to the 'big medical world' full time. I got tired of the twenty-four-hour, on-call basis and the hectic pace. Now, I call my own shots, so to speak, and I have much more time for designing and building suppressors. That's paying off very well, I might add," Dater says.

Success is not without pain, though, creating some very real problems for Dater which are reflective of all small businesses. When he's busy, he gets into the bind of not being able to offer fast service, yet hiring employees would tie him into more and more governmental paperwork.

"I'm up to my neck in paper now, just with my present business. If I add employees, I'll be spending more than 30 percent of my time handling paper on wage taxes, Social Security, unemployment, nondiscrimination records, etc. I'll be damned if I'm going to work as a free clerk for the U.S. government," he declares.

"I tend to get off on tangents when I get talking about supposedly free enterprise and big government. But now, I stick to what I know best and enjoy most—my company and my sound suppressors."

As he moved from custom work to a larger production and marketing status, Dater tried to simplify and standardize his line. He says, "I settled in on my most popular models and designs then standardized for production. This not only reduced costs, it also allowed for some shelf inventory."

Under the roof of his Automatic Weapons Company, Dater produces both integral and muzzle attachment suppressors for a wide variety of weapons. His integral models are more efficient, require some weapon modifications, and are sometimes more costly, while the ones which attach to the end of the muzzle are not as efficient, do not require weapon modification, and cost less.

Dater describes his integral units as "barrel suppressors built around the barrel porting system and using a muzzle suppression component. In my 9mm unit, for example, I use baffles in the front section and wrap screening around the ported barrel in the rear. I did use fiberglas, but have now gone to stamped baffles, instead. It just works better."

Dater likes this baffle and screen concept, saying, "not only does it tend to cool the gases, but it also diffuses them so that a shock wave doesn't hit the outer tube. Basically, I use a special aluminum screening because of high specific heat of the aluminum rapidly absorbing heat from the gases.

"Also, when I use the stamped baffles, I find that the more I use, within reason, the more effective the unit becomes. For example, in my earlier .223 and .30 caliber suppressors, I used eleven baffles. Now, I use twenty."

One of Dater's most popular designs is his RST, using the Standard Ruger .22 caliber pistol with a rather common, but highly effective integral suppressor arrangement. It uses the barrel porting principle with two suppression chambers. He offers the same arrangement with the Ruger MkI pistol for the more serious shooter.

"I've also altered the sights. My unit adds about three inches to the weapon and quiets it by a measured 27 dB during lab testing. Empirical observation shows a comparison with a CO2 pellet pistol in terms of muzzle signature.

"One of the benefits of this design is that the customer can do his own service—cleaning and repacking. The customer can buy all the materials he needs for about two dollars at his local hardware, then repacking takes about an hour. I tell people to repack after 500 to 800 rounds," Dater notes.

The overall suppressor length is six inches with a one-inch outer tube diameter. The weight of the suppressor component is five ounces.

He also offers his R10 for the Ruger 10/22 and an R11 for other .22 caliber semiautomatic and bolt action rifles. Both units are integral models, using ported barrels. His AR7 is, of course, meant for the Armalite AR7 Explorer rifle. These three units are essentially physically identical. Because of the barrel alteration, an accessory optical or telescopic sight must be employed. Dater's lab tests indicate a sound reduction level of 29 dB, providing about the same report as that of a Daisy BB gun. Fitted to a bolt-action weapon, the R11 reduces sound by a measurable 31 dB. Unlike some of his other suppressors, these require repacking after 2,000 to 4,000 rounds, which the customer can easily and cheaply do himself. The R10, 11 and 7 series suppressors are 14 inches in length, 1.25 inches in diameter, and each weighs sixteen ounces.

One design change instituted by Dater both simplifies production and makes his suppressors considerably cheaper and easier to own. He has

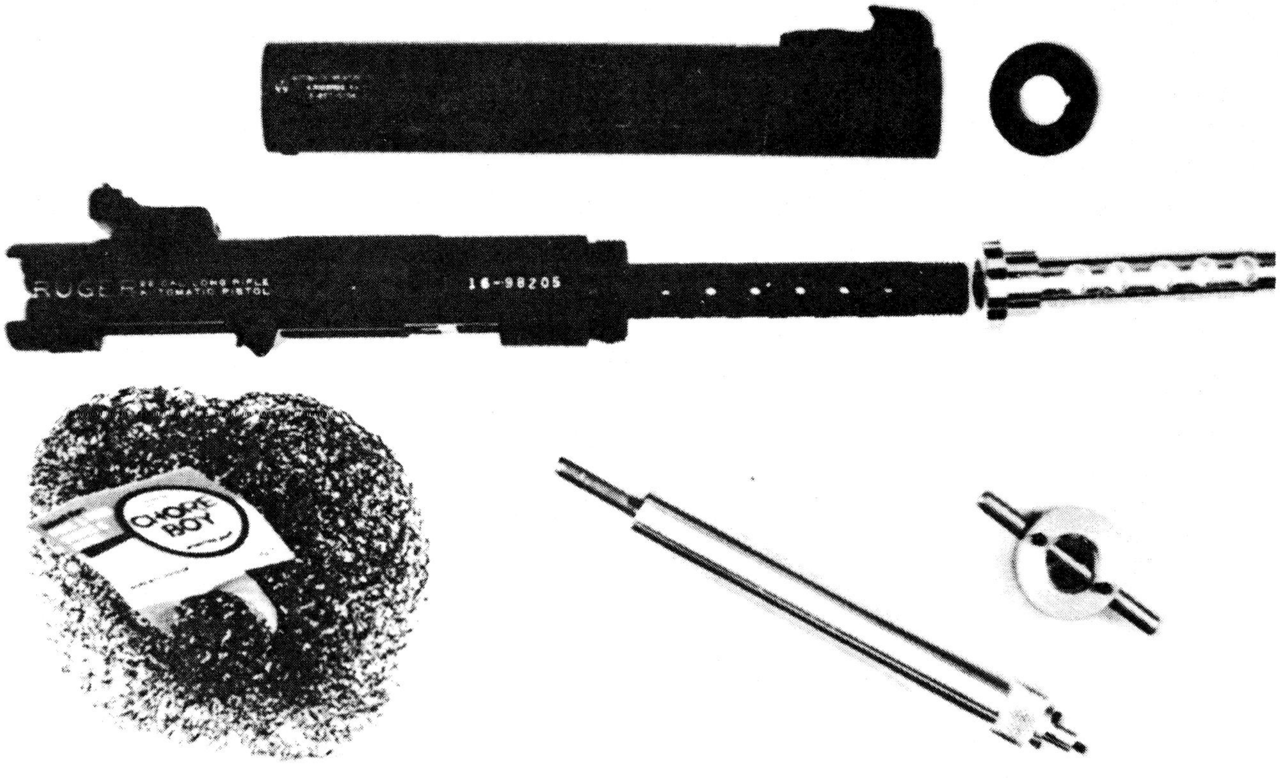

PHIL DATER

A totally disassembled RST suppressor with all parts, tools, and packing materials.

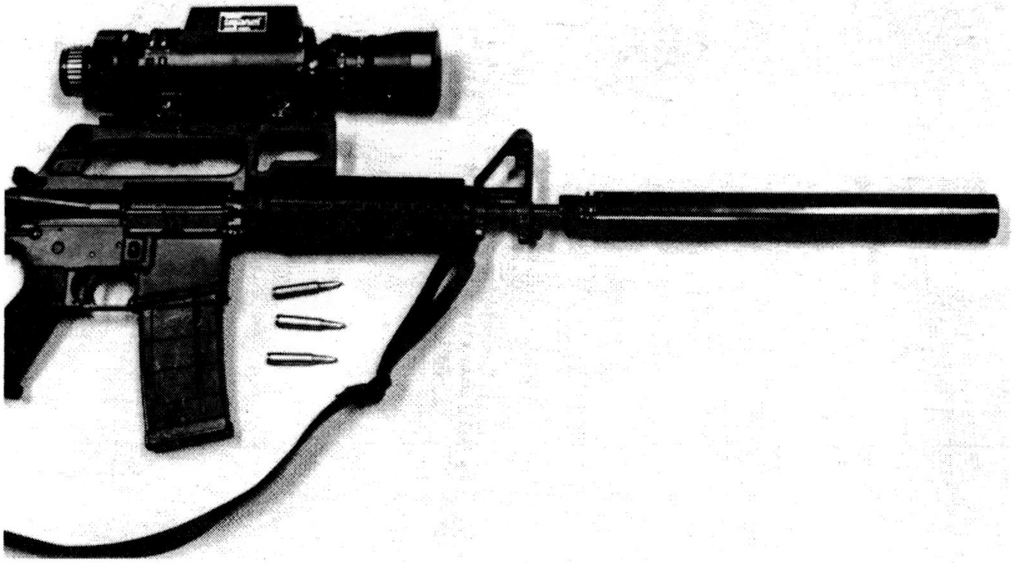

PHIL DATER

A Dater M22 suppressor mounted on an M16 with a Javelin night vision device.

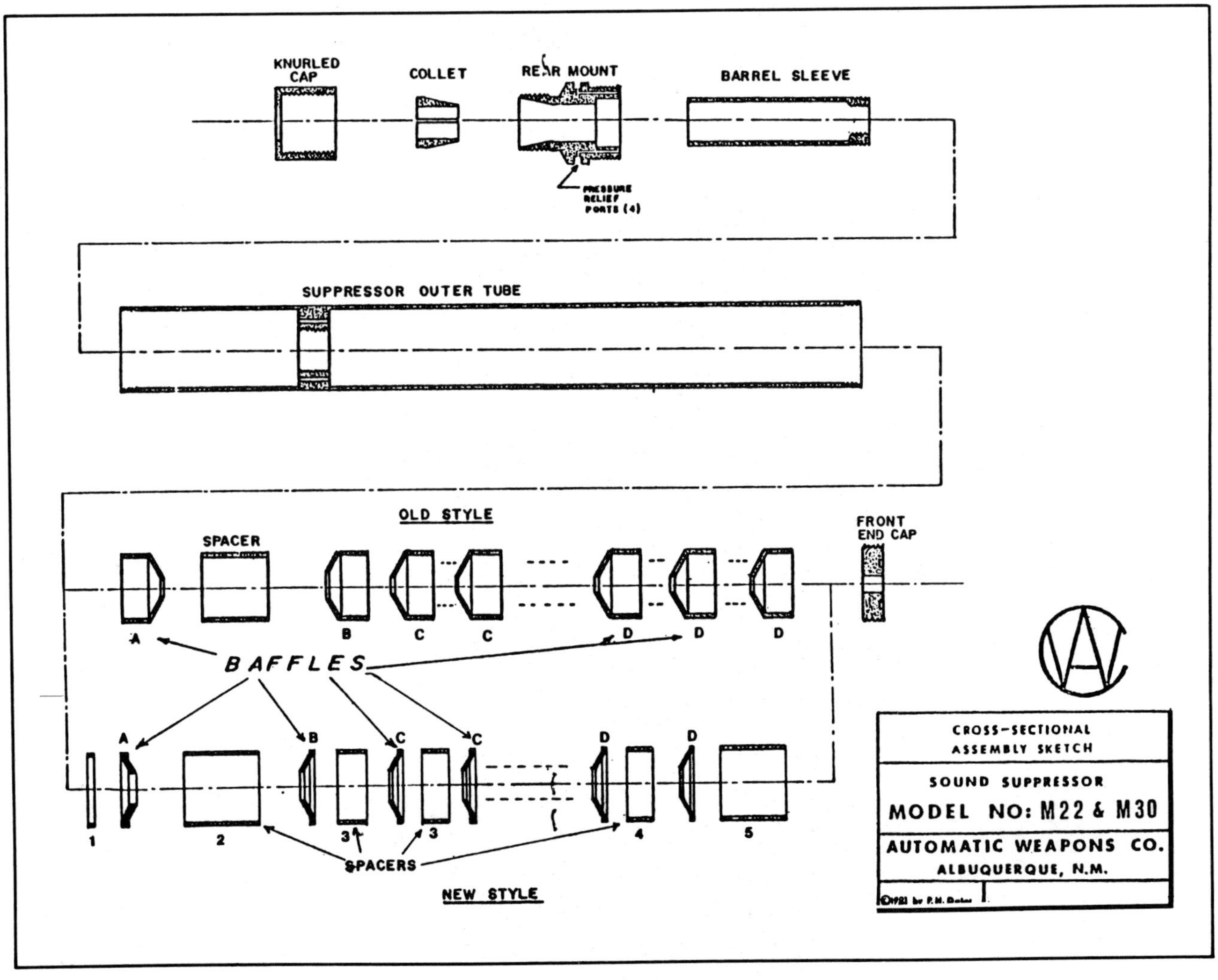

PHIL DATER

Cross sectional of the Dater M22 and M30 suppressors.

The model SG9 mounted on a Sten MKII SMG.

PHIL DATER

combined all his various 9mm integral suppressor models into one basic unit which he calls the SG9. His concept is aimed toward the individual who has a $200 NFA tax stamp on his one suppressor, but who also owns several 9mm submachine guns.

Dater explains, "Since the outer tube is the part that's serially numbered and registered as the suppressor, I make the one basic SG9 model for all 9mm weapons. Then I insert the correct barrel and barrel mount into the rear of the can to adapt to the customer's particular weapon.

"That way, if a customer purchases one basic SG9, he can use it for his S & W 76, say, then using only two adapter parts, he can use it on a Sten, or using another set of two parts, it will fit the MAC 9mm. All we change are the two adapter parts that mount the suppressor to the weapon."

This is the same idea that Dater uses with his M22 muzzle silencer, in which changing two minor parts permits adaptation to a wide variety of .223 weapons, e.g., M16, HK93, Mini 14, etc. The SG9 is 12 inches long and 1.6 inches in diameter. It weighs 2.3 pounds.

At press time, Dater had not yet had the SG9 measured scientifically on a laboratory machine, but from other measurements, it achieves sound reduction of about 21 dB, which is a very conservative measure. One knowledgeable military man who has tried Dater's SG9 on both a MAC10 and on an S & W M76, says, "Getting down 18 dB on the blast from a full auto weapon is damn good. Actually, his models we tested did a bit better—more like 23 dB."

Somewhat surprised, Dater admits, "I understate what my cans will do so the customer gets a little more than he bargained for. I'd rather do that than make claims my products can't live up to."

Dater also engineered an adjustment provision for the varying types of 9mm ammo that users are likely to run across. This allows the user to keep

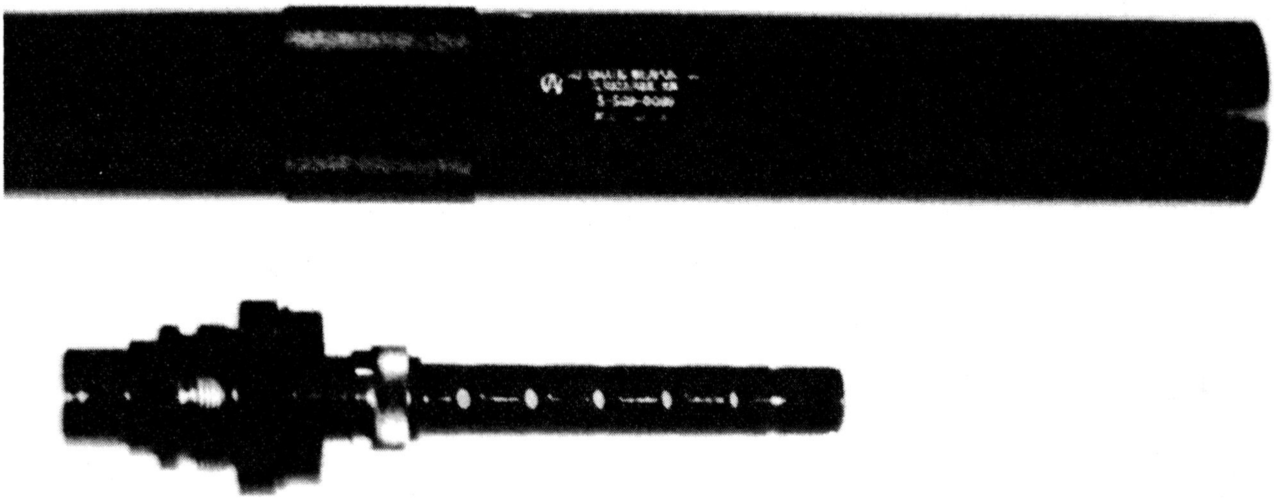

PHIL DATER

Phil Dater's SG9 suppressor with adapter for the Sten. This same basic suppressor may be used with a variety of 9mm weapons.

the ammunition near the subsonic level, yet still properly operate the weapon's action.

"I refer to this as my port-blocking collar," Dater says. "Some 9mm ammunition, especially reloads and some of the commercially marketed subsonic ammunition, may not have enough oomph to operate the weapon's recoil system properly. This is why I include a barrel sleeve, or port-blocking collar as I call it, to seal off the rear set of barrel ports.

"This increases velocity by delaying gas bleed-off in the rear of the suppressor. Of all ammunition, 9mm Parabellum is unbelievably variable in its loadings. Sometimes this can be a real headache, especially for a suppressed automatic weapon. But this collar overcomes that nicely on the SG9."

Dater's muzzle-attaching suppressors create another set of engineering problems, although not quite so complicated as those of the integral models. He notes, "Basically, you have to deal with muzzle and suppressor alignment, plus the addition of new optical or scope sights on some weapons because of the can's diameter. Also, you no longer have the advantage of a ported barrel, so subsonic ammunition must be used if you wish to avoid ballistic crack. That can create problems for the recoil systems of some weapons. But, with other weapons and systems, a screw-on suppressor works wonderfully."

Dater's muzzle suppressors include his R22 in both regular and slimline models. These screw-on suppressors are designed for a variety of .22 rifles and pistols. Both models are 6.5 inches long, while the standard has a diameter of 1.25 inches, as opposed to the slimline's 1-inch diameter. The advantage of the slimline is that the weapon's normal sights can be used in most instances. The slimline also uses internal baffles rather than the usual replaceable packing materials. Cleaning the slimline is done by immersion in a suitable solvent.

The degree of suppression of these units is 25 dB, as measured using proper equipment. An absolute measurement is difficult because, as Dater points out, "The absolute sound level will vary significantly with the type of weapon and barrel length. However, the degree of sound reduction is quite suitable for undetected backyard shooting in residential areas at, say, garden pests."

Phil Dater's port blocking collar for his SG9 suppressor shown dismounted, then mounted on a S & W 76.

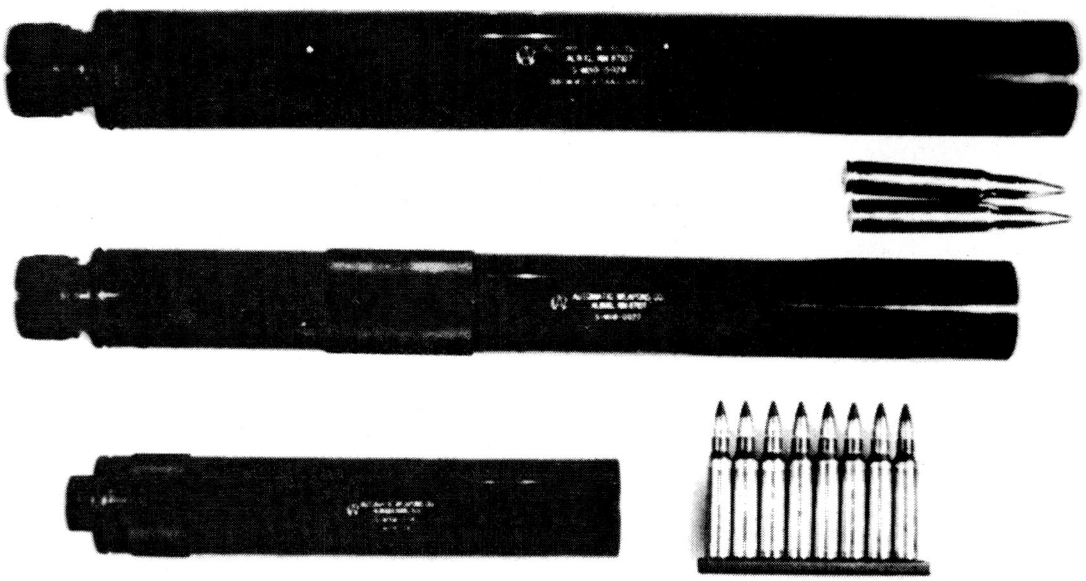

PHIL DATER

Dater also markets a line of centerfire muzzle suppressors. Top is the M30 in .30-06; middle is the M22 in .223; bottom is the MXM in .223.

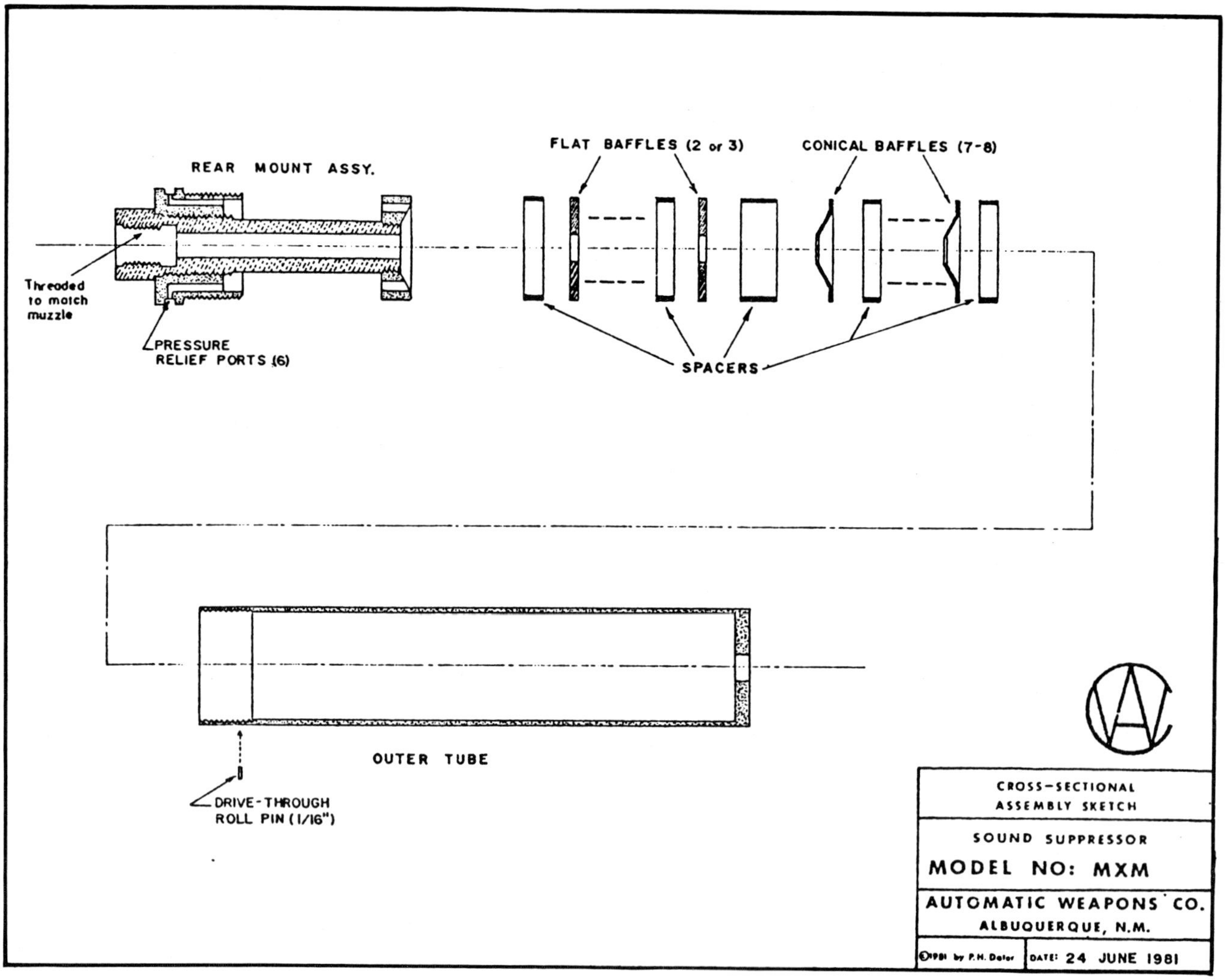

A cross sectional sketch of Phil Dater's MXM suppressor.

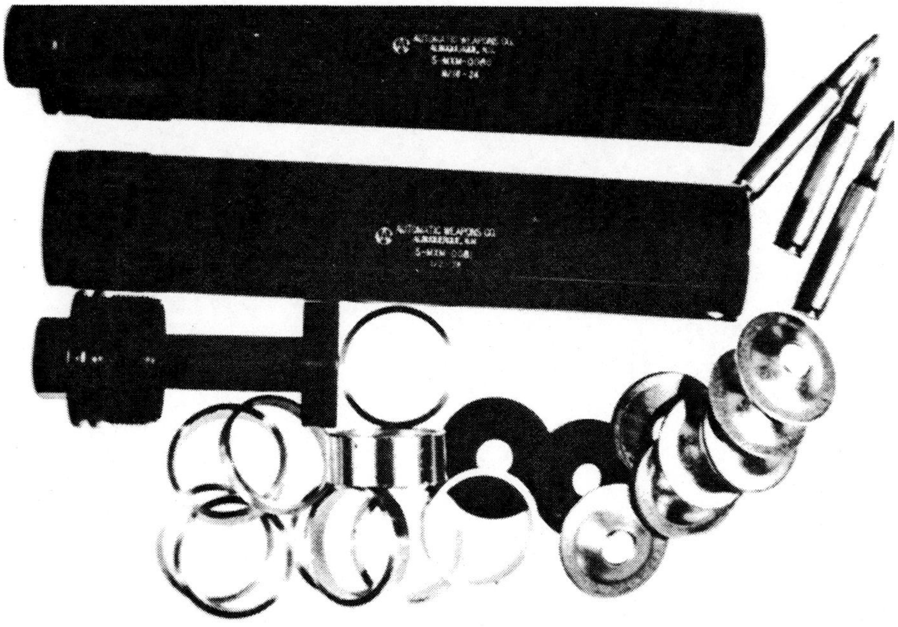

The disassembled MXM suppressor.

PHIL DATER

A totally disassembled Dater model TC3 suppressor in .357 magnum for the Thompson centerfire pistol pictured—one of Dater's custom designs for this powerful handgun.

PHIL DATER

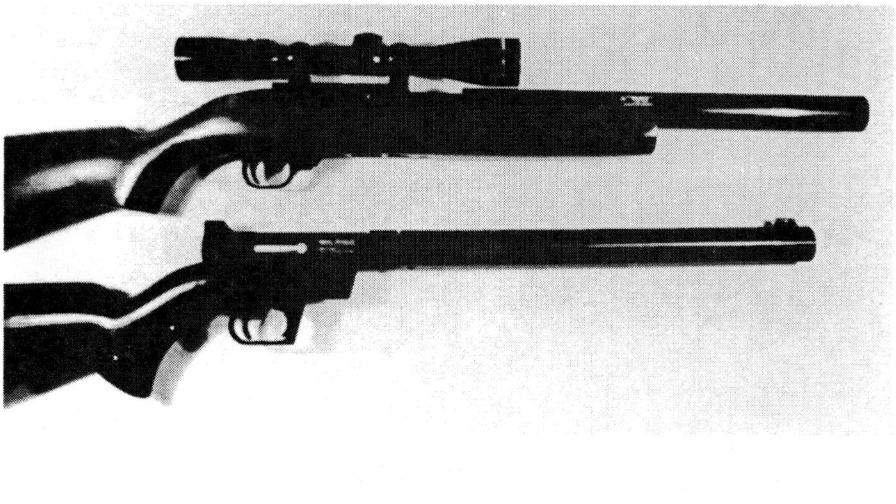

Phil Dater's R10 suppressor on a Ruger 10/22 (top). Bottom is his AR7 suppressor on a Charter Arms AR7.

PHIL DATER

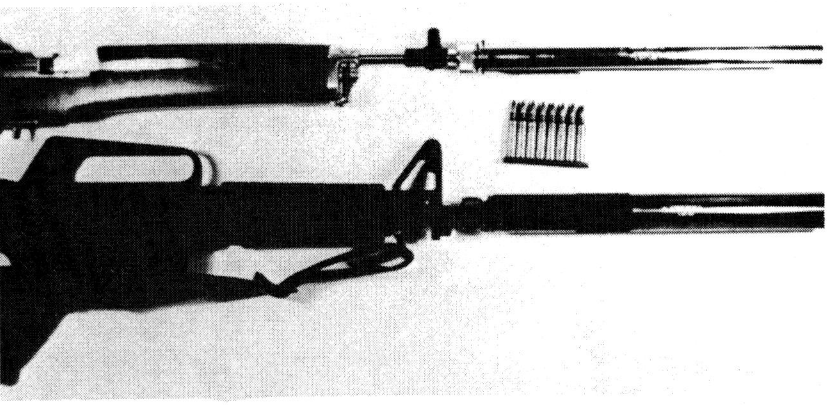
PHIL DATER

Phil Dater's M22 suppressors. Top is a stainless steel, M22-SS on a stainless Ruger Mini 14. Bottom is the standard M22 on an M16.

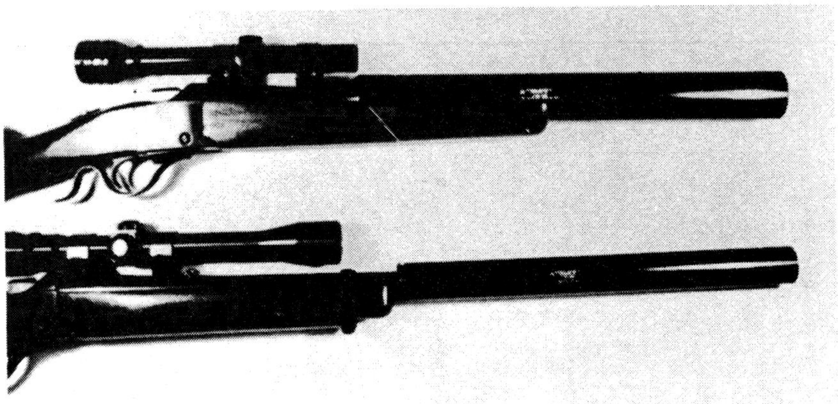
PHIL DATER

Dater also does custom suppressors. At top is an R3A custom unit on a Ruger No. 3 in .45-70, an integral unit. Bottom is a .223 muzzle suppressor.

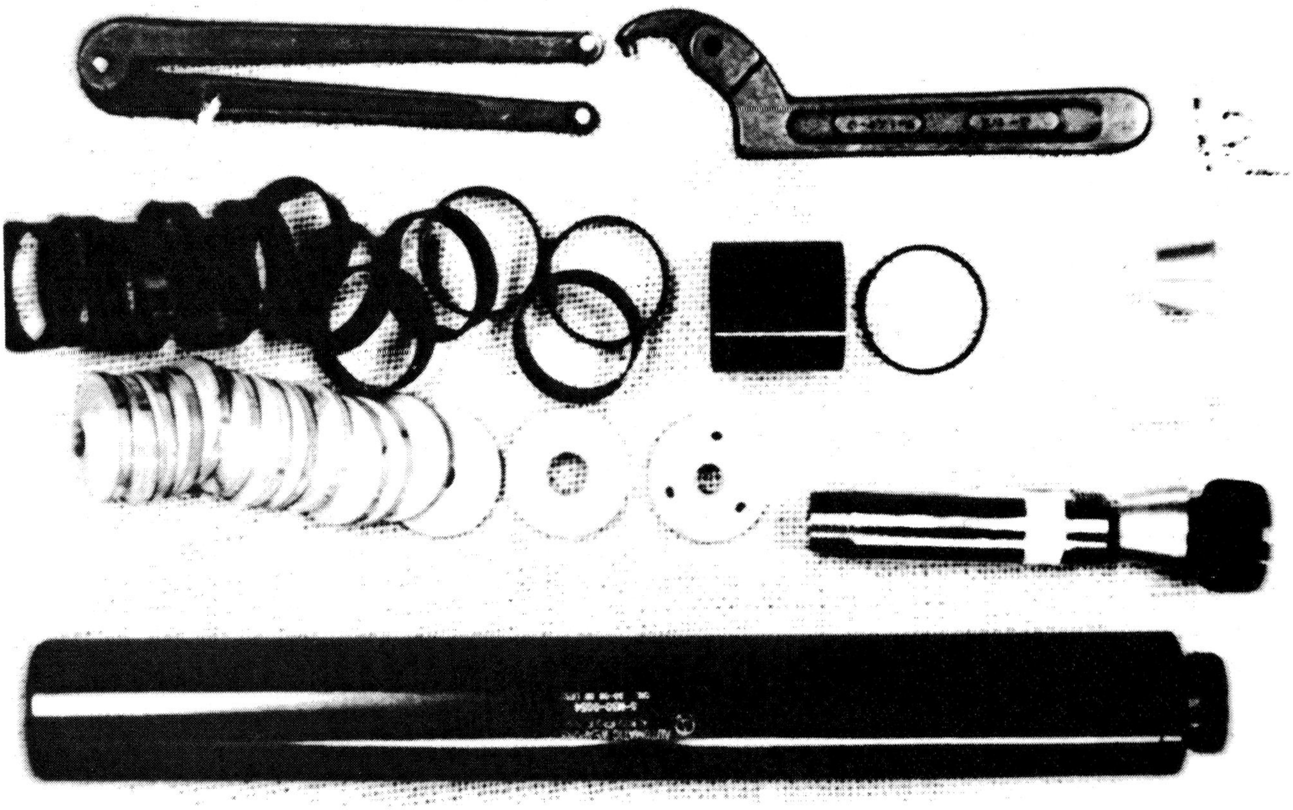
PHIL DATER

The M30 totally disassembled. This early model uses commercially purchased spanner wrenches. Dater now builds his own tools for maintenance.

Dater places much of his design emphasis and testing on empirical, subjective judgment about the efficiency and effectiveness of a suppressor. He says, "There is probably too much emphasis placed on the scientific, laboratory approach, especially when field conditions are ignored. People don't use a suppressed weapon in the lab, they use it in the real world.

"However, I asked Reed Knight, a suppressor-knowledgeable expert, to measure my unit. His test jig is a Ruger pistol with a two-inch barrel. I was pleased with the conservative minimum 25 dB drop."

As a new wrinkle, Dater has begun using a computer as an aid in designing and improving his suppressors. He says, "I have my program almost bug-free at this point. The main thing the computer does is factor in a lot of science and numbers, suggesting suppressor size, baffle numbers and placement in given caliber, barrel length and powder charge/bullet used. These variables seem to be prime considerations in determining gas volume, and I know I am on the right track, as the lower test numbers are there."

He realizes, though, that like the EPA's gas mileage figures for automobiles, the ubiquitous and lab-spawned dB ratings are essential to the commercial comparison of the many available suppressors.

He completed a series of lab tests on several of his screw-on suppressors as this chapter was being written. Dater's instrument measurements are taken at ninety degrees to the side and at five meters distance, except for the .223 and .30-06, when he moved out to ten meters. On his custom design for a Ruger #3 in .223, Dater achieved a 26 dB drop. His custom prototype for a .300 Winchester magnum showed a drop of 24 dB. His prototype M22 suppressor with fifteen baffles, mounted on an M16, showed a drop of 25 dB. A custom can on a .223 Sako cuts sound 28 dB.

"I would say that at seventy-five meters my M22 unit is essentially inaudible. If you are using one on an M16, you can hear only the weapon mechanism working and hear the round impact seventy-five meters downrange."

The production M22 is designed as a screw-on for a variety of centerfire .22 rifles, including the AR15, M16, AR18 and AR180. With an accessory internal barrel sleeve, it can also be used on the Ruger Mini 14, Ruger AC-556, the Heckler & Koch HK93, and a wide variety of other .223 weapons. The length of the M22 is 13.25 inches, the diameter is 1.375 inches, and the weight for the standard model is two pounds.

He also redesigned his SM-XM and renamed it the MXM. It's a sound moderator for the XM-177E2. Its mission impossible is to hold down the awesome muzzle blast that goes along with the mighty weapon. "I used stamped baffles in it with some other design changes, and, amazingly, I achieved a 19 dB reduction in sound level," Dater relates.

Basically, his MXM is a markedly shortened M22 unit with a single point mount on the muzzle threads. The finished unit is 7.5 inches long and 1.38 inches in diameter.

The R22Mg is an inexpensive muzzle suppressor designed for the American 180 submachine gun. It is not as effective as the integral type AM8 suppressor which Dater custom-designed for the American 180, however.

Dater has now adopted a small .22 to the Walther PPK in .22 caliber. He says, "I made up a few of what I call my 'James Bond units' for some customers. The unit is not quite as effective as the can for the Ruger, but close. It makes a good backpacking weapon, by the way."

He is also doing the same basic design for the AMT .22 Backup, using a stainless steel suppressor with a sound level drop that tests at 25 dB.

In addition to his standard items, Dater will custom design and manufacture suppressors for almost any other feasible weapon. He has a scaled-up version of his M22 for various .30 caliber rifles. It's the M30 which is 14 inches long, 1.6 inches in diameter, and weighs 4.5 pounds. He also produces, on a semicustom basis, suppressors for the M3A1 submachine gun, the American 180, the Ruger #1 and #3 rifles, Thompson Contender models, and various other weapons.

Dater's order file really began bulging after *Soldier of Fortune* magazine ran a colorful article alliteratively titled "Doc Dater's Deadly Devices." Dater smiles and says, "Yeah, the writer who did the piece was so impressed that he ordered some units for himself, and apparently the word went around the circuit that folks who wanted one of my suppressors had better get their orders in early. It was quite a pickup for my ego and my business ledger."

A later article in *SWAT* magazine, which Dater

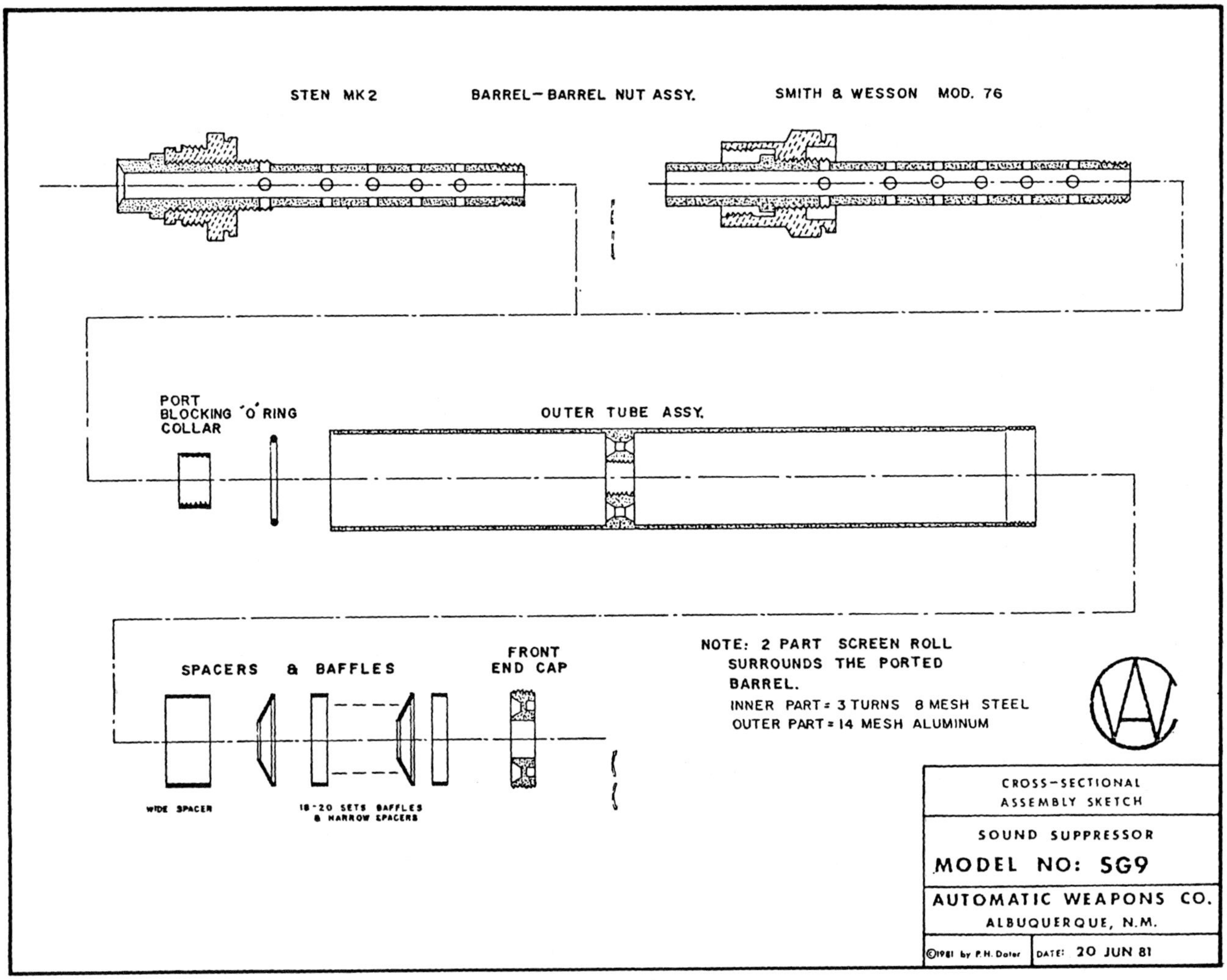

PHIL DATER

The cross sectional sketch for Phil Dater's newly improved suppressor model SG9.

PHIL DATER

The disassembled M22 suppressor.

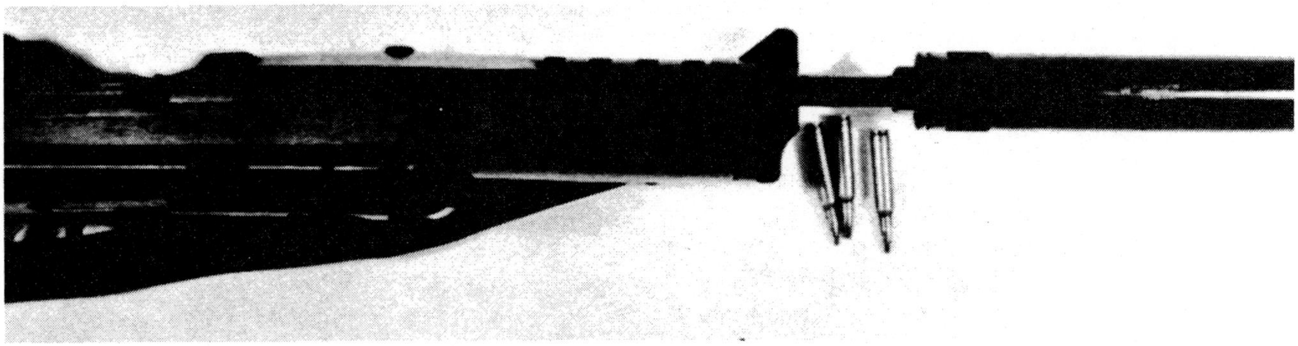

PHIL DATER

The model MXM suppressor mounted on a Ruger AC-665K, full auto weapon in .223.

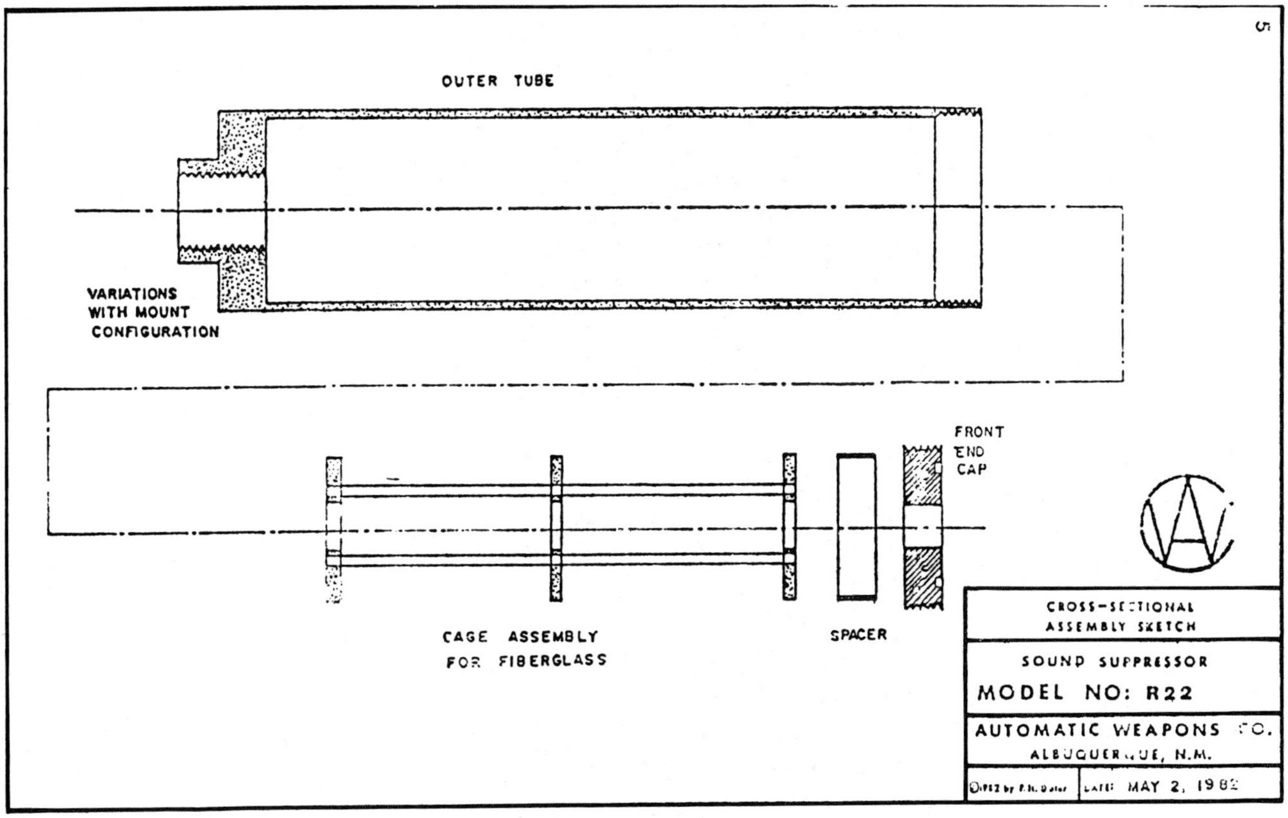

PHIL DATER

Diagram for the R22 Automatic Weapons Company as designed by Phil Dater.

wrote himself, also brought in many more inquiries and sales. He says, "My business is booming. They seem to like my stainless steel units."

Asked about his business philosophy, Dater says, "I follow the advice of both John Ruskin and W. C. Coleman. According to Coleman no product is really sold until it is delivering satisfactory service to the user."

What Ruskin said, and what Phil Dater uses as his working policy about pricing his products, is as follows:

> It is unwise to pay too much . . . but it's worse to pay too little. When you pay too much, you lose a little money . . . that is all. When you pay too little, you sometimes lose everything because the thing you bought was incapable of doing the thing it was bought to do. The common law of business balance prohibits paying a little and getting a lot. It can't be done.
>
> –John Ruskin

As of January 1983, his prices ranged from $155.95 for an R22 model to $623.00 for the M30 unit. But prices are relative to the time, the demand, and the marketplace. As Phil Dater quips, "Sometimes, my catalogs are out of date before I get them to the printer."

In response to the quest for an ideal, universal design, Dater just shakes his head. "Design of the universal can which could simply mount to *any* weapon would be the admirable goal in the annals of suppressordom," he says. "But, it would also be grossly impractical and inefficient, unless one were propmaster for a film company. Their ideal would be one inch in diameter, four or five inches long, and set with a universal mount to fit anything from a tiny .22 caliber hideaway, to Clint Eastwood's .44 magnum, or maybe a quad .50!

"Then the sound man would make a quiet 'phutt' sound and that would be it. No worries about weight, degree of suppression, gas volume, velocity, metallurgy, or those other gremlins which bug us guys in the real world.

"There is also some current thinking in the real design field with which I disagree," Dater says carefully. "For instance, I can't see anyone claiming they build units which never need to be serviced. Hey, I know how dirty any .22 gets. So does anyone who's fired one. Barrel ports fill up with bits of lead, power residue, and bullet lube. You have to perform some cleaning service.

"I also think it is wrong to use porting holes less than an eighth inch in size. Smaller than that is just too fine to deburr properly inside the barrel. I have seen such units show considerable lead shaving on the bullets."

In addition to his silencer work, Dater expanded his product line into electronics. He's designed and is producing an electronic counter for both the Dillon and Star ammunition loading machines.

"The device counts loaded rounds as they are ejected from the machine and also has a buzzer that sets off the alarm every 1,000 rounds, which is of benefit to the commercial reloader. Or I can use a relay which automatically shuts off an electrically operated press, such as an AmmoLoad," he explains.

Dater also markets his combination catalog/manual on suppressors. It's a very complete publication and is available from him for three dollars at Automatic Weapons Co., P.O. Box 1731, Socorro, NM 87801.

Talking about his business venture, Phil Dater says with the look of a very satisfied man, "I am spending many more hours at this per week than when I was practicing medicine full time. But, I am enjoying it and myself a lot more."

6. He Says He's Number One

He advertises himself as "the foremost manufacturer of suppressed firearms today . . . the standard of quality and performance the others try to match." He could be telling the truth.

Whether Jonathan Arthur Ciener really is number one isn't the point. The only people who count in judging that are the military and intelligence technocrats who evaluate his designs with standard issue adoption in mind, and they're not talking. So, only history will really determine if Jonathan Arthur Ciener was the number one silencer-maker of the 1980s. According to him, though, he is always concerned with his reputation and the quality of his product.

"All my designs are built around two major points—performance and long life. I strive to moderate the sound as much as possible without damaging the accuracy of the firearm. I am devoted to ensuring that a unit fires as well after suppression as it did before," Ciener notes.

"For as long as I can remember I've had an intense interest in firearms. I bought my first gun, a Browning .25 auto pistol, at age twelve. That's when I began my collection. This collection branched into military weapons, and then, where else, but into full auto weapons. I already had an M1 carbine, so I bought an M2 conversion kit, which I registered during the 1968 amnesty program. That's how I started."

Ciener credits his engineering and design abilities to working with his father and brothers as they repaired, rebuilt, and remachined automobiles and motorcycles. "I learned how to design, fabricate, improvise, improve, and machine all sorts of mechanical parts," he adds.

He both advertises and advocates quality and service life in his designs. Like his colleagues, Ciener recognizes the need to diversify his marketing to reach the various potential markets. Yet, he is also very practical. All silencer makers, from

JONATHAN ARTHUR CIENER

Ruger 10/22 Sporter with integral suppressor. Note the clean lines of integration of the suppressor with the frame. The unit is ultra quiet, has no muzzle wipe, and yet has good life expectancy.

JONATHAN ARTHUR CIENER

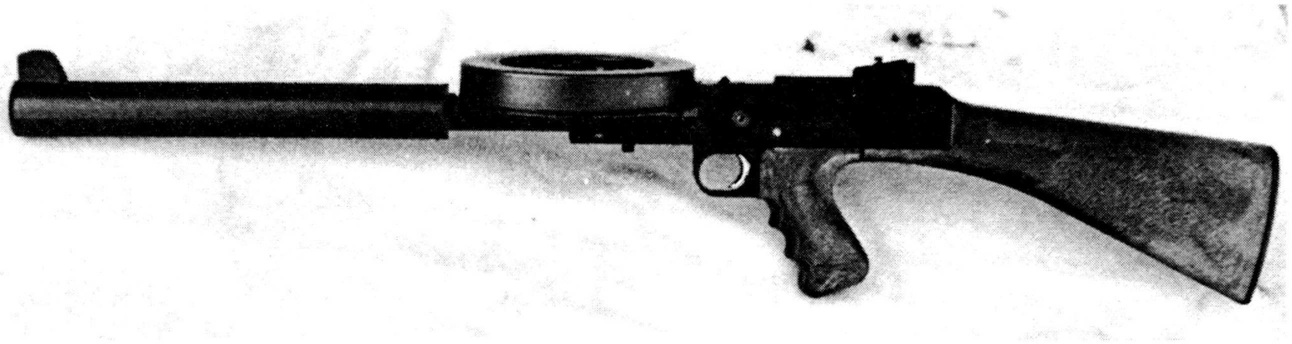

American 180 suppressor/barrel assembly.

JONATHAN ARTHUR CIENER

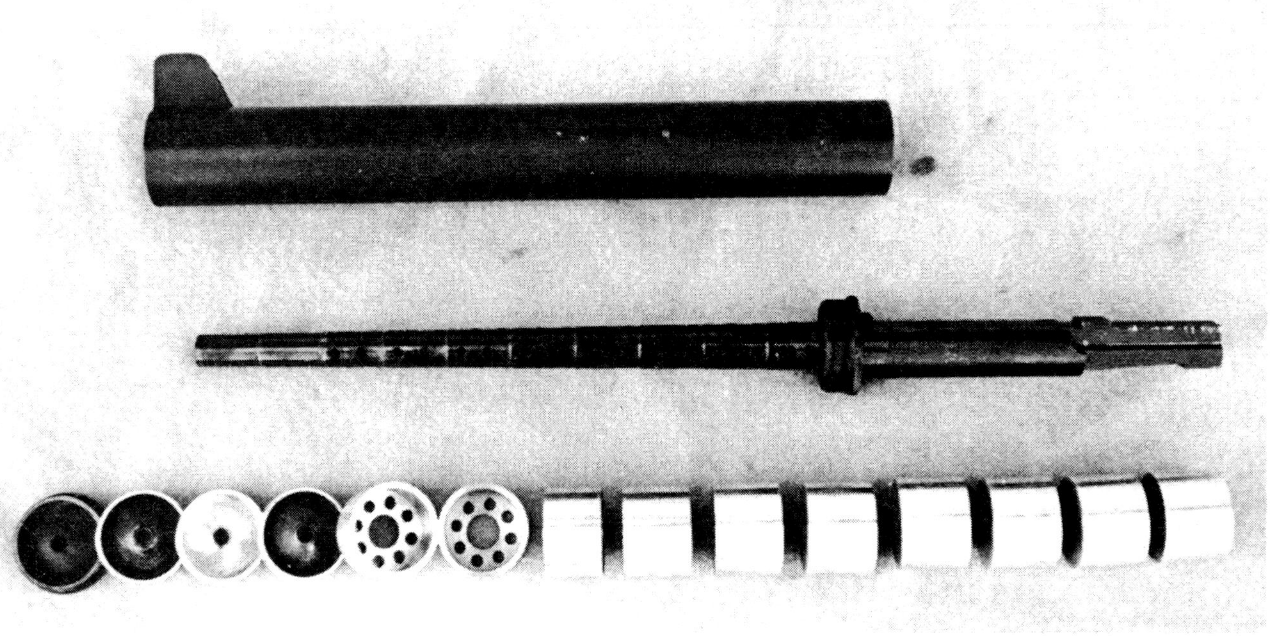

American 180 suppressor is only .22 cal. unit Ciener presently builds that can be disassembled for cleaning. He felt this was necessary due to high volume of fire put through this weapon. Precise machining necessary to insure that barrel drill holes line up with chambers of baffles.

JONATHAN ARTHUR CIENER

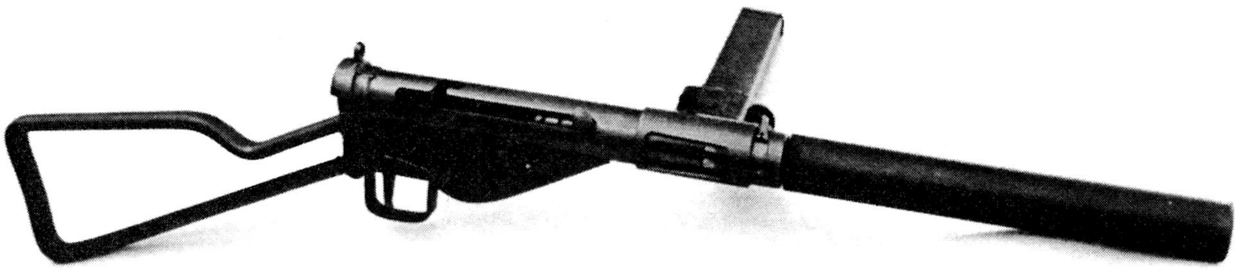

English Sten Mark II. Same features as Uzi, S & W 76, MAC 10 and M3/M3A1 units.

H.P. Maxim to Ciener and his peers, know that the only long-term customer relationship worth going after is the various military, police, and other governmental agencies with special needs.

Ciener says that despite official denials, suppressed weapons have been used in American covert actions since the Vietnam war. This revelation disputes official comments that "no special mission weapons or tactics have been employed, including suppressed weapons, since test limited field testing and evaluation and use in Southeast Asia," as James W. Dunne, a Defense Department employee, told me.

In May of 1981 Ciener said in a press interview that the Delta Team commandoes America sent in to rescue the hostages from their Iranian captors were armed with suppressor-equipped M3A1 submachine guns. Ciener said he could plainly see the weapons in the various news pictures showing our wrecked helicopters in the Iranian desert.

Like his competition, Ciener is aiming solidly for this type of market. For example, according to Ciener, his units are ideally suited to SWAT team marksmen.

"Concealment is a necessary part of his mission, as he must often maneuver into a position to fire from. After the sound of his first shot with an unsuppressed weapon, his position must be changed to avoid return fire. With our suppressed rifle, he is not put in danger, as the sound is nondirectional, giving the adversary no idea as to his location to return fire," Ciener says.

According to Ciener, the rest of the SWAT team finds that the use of suppressors on their weapons allows for superior command and control.

"Many a hostage situation rapidly deteriorates to an open gun battle with casualties because the noise of the first fired weapon provokes all those involved to fire unnecessarily," he adds.

"With a plan of extraction, as in the case of the Iranian embassy operation in England or our ill-fated Iranian mission, success depends on the team being able to move from adversary position to adversary position neutralizing them [the enemy] as they move but not alarming the next position they must assault. Otherwise the hostages may be harmed before the team can extract them. Suppressed weapons are a must."

Since most of these special mission teams use both semi- and fully automatic weapons, they are the prime area for Ciener's developments.

"Automatic weapons benefit from suppressors because they allow greater accuracy of fire by reducing the recoil. We have designed light-weight, efficient units for virtually all submachine guns as well as light belt-feed weapons," Ciener says.

JONATHAN ARTHUR CIENER

M14 with a Ciener suppressor. With bipod, scope and the outstanding Ciener unit, this is an excellent weapon.

JONATHAN ARTHUR CIENER

H & K unit is adaptation of M1A/M14 or AR15/M16 unit. Ciener can adapt this basic design to FN-FAL, Remington 700 or almost any bottle neck cartridge gun.

JONATHAN ARTHUR CIENER

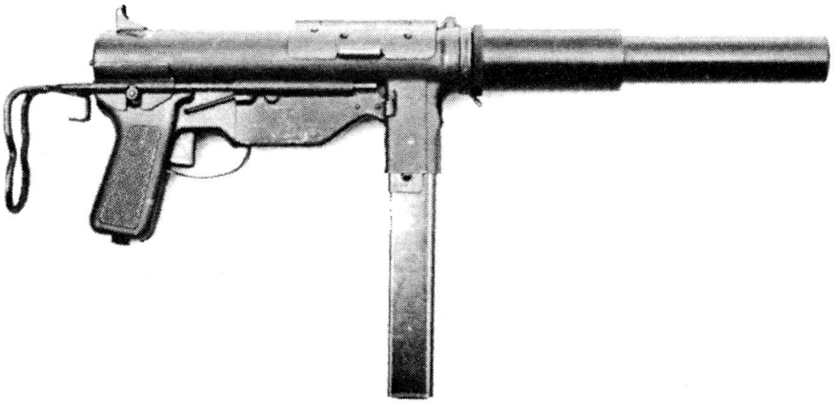

M3/M3A1 suppressor barrel assembly.

JONATHAN ARTHUR CIENER

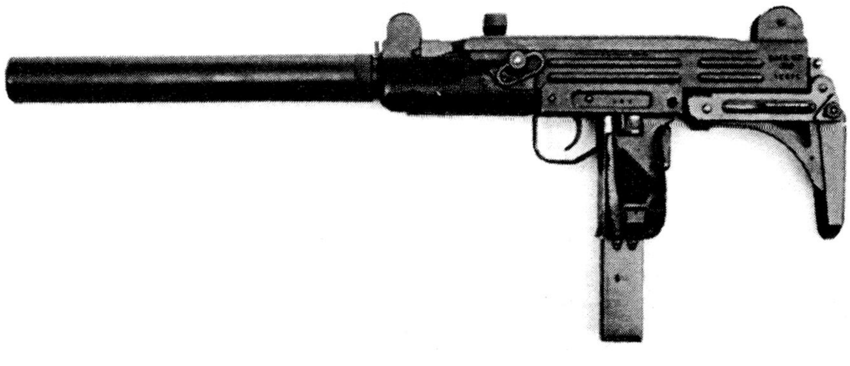

Uzi suppressor/barrel assembly has features of S & W 76, Sten, M3A1, etc.

JONATHAN ARTHUR CIENER

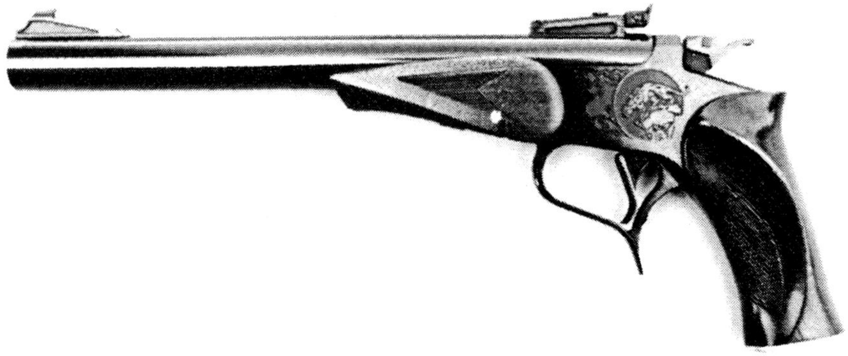

T/C Contender .22 caliber. An example of specialized units by Ciener. Ciener can and does supply all Contender barrel designs and calibers upon request. Most are readily disassembled for cleaning. Others don't need it. Superior firearms for suppression since they are single shot, with no noise lost through action upon functioning.

JONATHAN ARTHUR CIENER

Walther PPKS .380/.22 with screw off suppressor. Interchangeable barrel that is three inches longer than standard allows Ciener to use bleed off holes and holding chamber as in H & K's very successful MPS/SD. This weapon has the additional feature of easy disassembly of suppressor for cleaning. No wipes to wear out or affect accuracy. Virtually all noise escaping from unit is from ejection port upon functioning. Optional slide lock improves performance as in all auto pistol designs.

As his governmental business solidifies, Ciener is standardizing his stock line of weapons to include the M16A1, Ruger MkI, M1A/M14, Remington 700, Ruger 10/22, Browning .22 auto rifle, and the Marlin 780, which Ciener describes as "virtually silent." He also markets barrel/suppressor assemblies for the S & W 76, Uzi, Sten MkII, M3A1, and Thompson Contender. There are certain weapons which either lend themselves well to suppressors or are popular with the special mission people who use them. For this reason, standardization is relatively easy. However, Ciener will also do any custom work necessary for any legitimate customer.

"We do custom designing and fitting of our suppressors on just about any other weapon the customer has in mind. Be assured that we have or can procure almost any weapon you may wish to own," he adds.

One of his more ambitious designs was a unique suppressor for the American 180. He recalls, "That 170 round magazine and the 1200 rmp cyclic rate requires a very special suppressor design. Most existing units would be worthless and would need rebuilding after only two magazines of operation.

"But, not mine," he adds with pride. "Our easily disassembled unit was designed with this specific weapon and its user in mind. It's rugged, long-lasting, and it will do the job. Our barrel/suppressor assembly can easily be disassembled, cleaned, and assembled in minutes."

Like others, Ciener also sees a great deal of value in the military using suppressed weapons to introduce reticent trainees to the firing range.

He notes, "Suppressors make people who are fearful of firearms more at ease when they're around them. I have personal experiences in which I have gotten nonshooters to fire weapons and become avid fans when we used a suppressed weapon. There is a very real difference between firing weapons with ear covers and firing the same weapon with a suppressor. No worry about flash, blast pressure or comfort while shooting are some of the advantages to suppressed weapons."

Ciener is also very familiar with the technology of firearms suppressors. Noting the difference in caliber pressures, for example, he blends his design techniques and the materials used to match the needs of the smaller calibers as well as the larger weapons, all while keeping the physical configuration of the units practical for field use.

He explains, "Lower caliber firearms can use lighter-strength materials. We have found good

JONATHAN ARTHUR CIENER

M1A/M14 suppressor. Same features as AR15/M16 unit.

American 180 suppressor/barrel assembly with laser sight mounted.

JONATHON ARTHUR CIENER

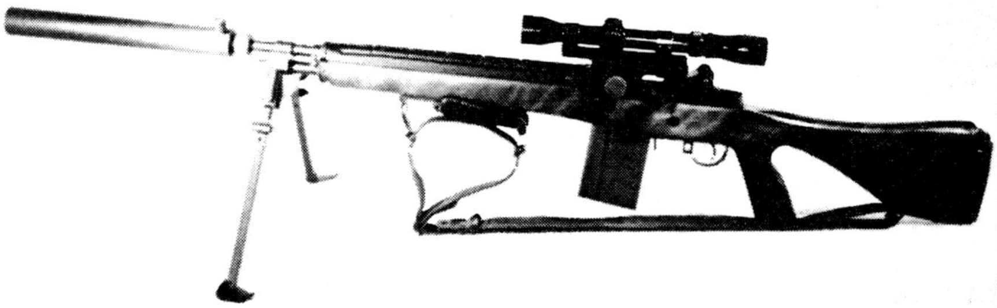

The Ciener suppressor on an M14 with bipod.

JONATHON ARTHUR CIENER

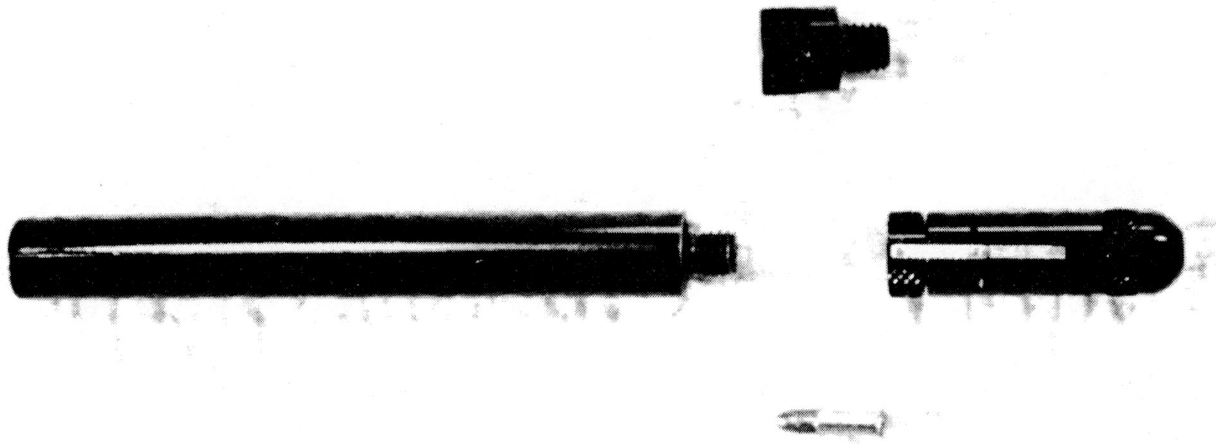
JONATHAN ARTHUR CIENER

An original MAC Stinger with a Ciener suppressor, sound quieting better than the original.

results with light, porous materials to slow and cool the expanding gases. Porosity is necessary to allow the lead and powder residue to dissipate.

"Due to the high pressure associated with them, larger caliber suppressors require materials that sustain higher energy levels. Suppressor designs for these calibers must take this into account. Here it is necessary to capture the energy pulse and use it against itself with the confines of the suppressor. To do this, we reverse the direction of the pressure pulse, thus much of this pressure can be "trapped" in a holding chamber and released gradually as the pressure in the recirculation area lowers."

Ciener adds, "This technique requires the greatest precision in machining tolerances, as the center hole the bullet passes through needs to be as close to the bullet diameter as practicable. This traps the gases behind the bullet forcing it into the suppressor design."

Where this suppressor designer is concerned, the machine tolerances aren't the only critical element. Ciener is as critical of the work of others in his field as he is exacting with his own. He says of suppressor technology that "while the basic technology of firearms suppression has been around now for seventy-plus years, recent developments have come in three forms: new internal materials; recombination of known techniques; and gimmickry."

He speaks with disdain of the last one, saying, "Gimmickry is apparent to me in some of the most publicized recent designs. The spiral diffuser that has enjoyed recent popularity is one example of this. The speed at which the pressures inside a suppressor move most certainly counters the theory of forcing the gases against each other. The spiral design would only work in a longitudinal direction, i.e. front to rear and then back. What is needed is a swirling motion to keep the pressure waves directed away from the muzzle outlet.

"The second gimmick is the so-called pressure relief valve. This consists merely of an outlet in the suppressor that has some holes in it to restrict the exit of the gases to the atmosphere. To be effective, it must have some form of retention system until the pressure is reduced.

"In our tests the same unit without a relief valve was quieter than with and a unit fitted with an extra valve was louder. Also, in our night tests, there was a definite fireball at the valve."

A third gimmick which Ciener feels is less than useless is the muzzle wipe, a bit of 1960s silencer technology in the form of a rubber disc whose function is to close the center hole of the unit after the bullet passes by, thus shutting off escape of gases.

"The much-believed idea that this wipe has any effect on the velocity of the projectile has yet to be demonstrated to my satisfaction," he says. "Actually, the wipe has been shown to be detrimental to the accuracy of the weapon. It rapidly deteriorates as the suppressor is used, which then deflects the trajectory of the projectile. This is of

Jonathon Ciener with a Stoner conversion, firing the Ciener suppressor.

Always a popular weapon, the Israeli Uzi goes very quietly with the Ciener suppressor.

little concern in full auto fire, but in the semi-auto mode, accuracy will be very poor, making an 'open bolt' firing weapon, already inherently terrible for accuracy, even more so.

"In high-velocity, long-range shooting, any contact with the bullet after it leaves the rifling will affect the accuracy adversely. Trajectory is so adversely affected as to render the rifle worthless. So, as the exit hole in the muzzle wipe is rapidly enlarged through wear, any advantage in sound suppression is quickly diminished."

Yet, this outspoken designer is impressed with much of the new technology and feels that space-age thinking and materials will be the next benchmark in silencerology.

He says, "New space-age, nonresilient rubber, nylon, and plastic derivatives that have enhanced sound absorption and heat retention are coming forth from our scientists in the space program. As with other industries, we suppressor designers need to apply these new developments to our needs.

"The materials we currently use have the drawback of limited life—they rapidly deteriorate. For military applications, this is acceptable. As the effectiveness diminishes, after a certain number of shots the individual military user has an insert pack that brings the unit back to its original effectiveness. Depending on the ammunition used and rapidity of fire, this can be as little as five shots. For commercial sales applications, this is not as favorable.

"Who knows what materials exist or are soon to develop that will both be effective and offer

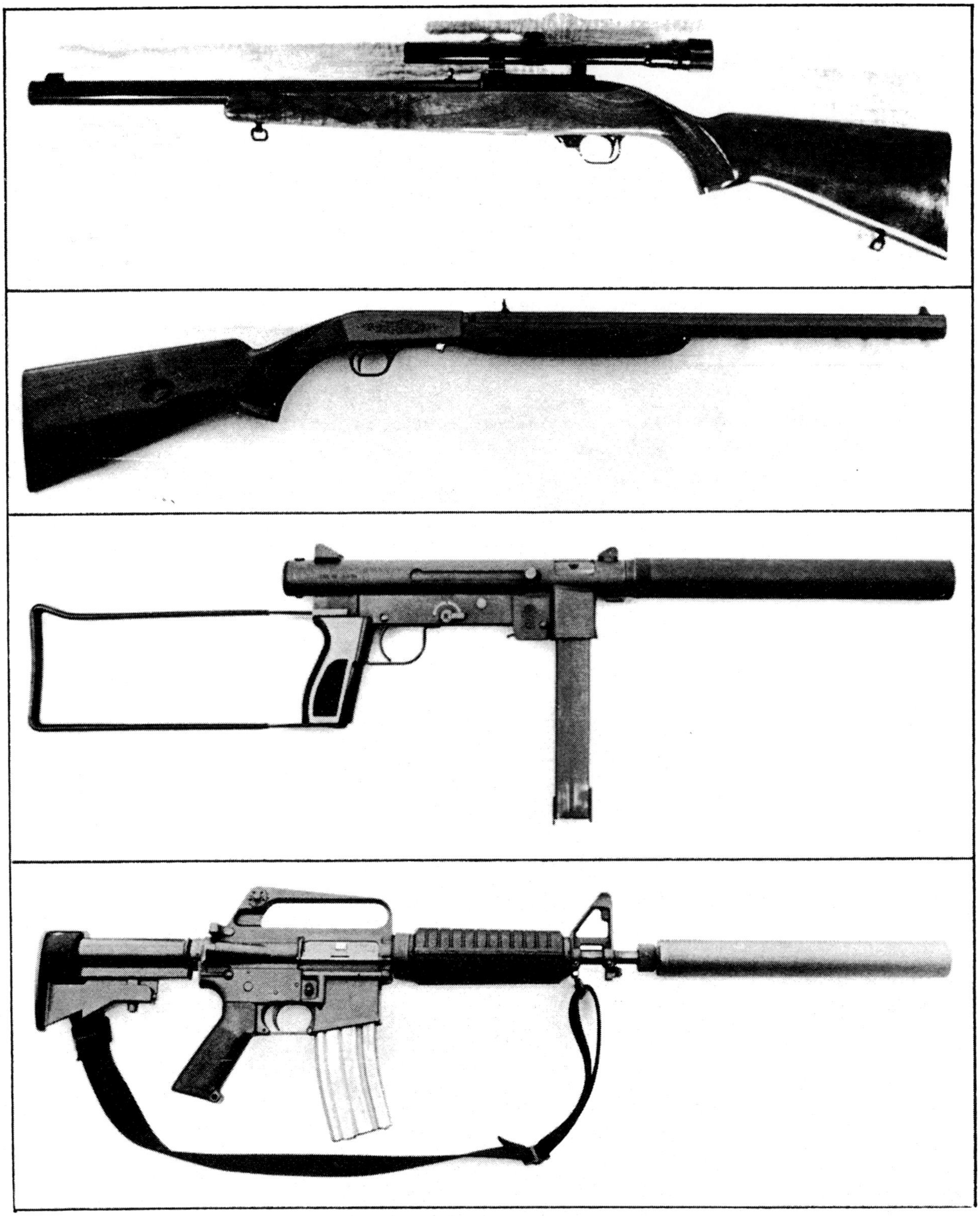

JONATHON ARTHUR CIENER

(Top to bottom) 1) The clean, streamlined, dependable Ruger 10/22 is one of the finest rifles in its genre today. Adding the effective, beautifully finished Ciener suppressor adds to the value of the weapon. 2) Browning .22 has all the features of 10/22 Sporter, but this gun has downward ejection so noise loss is directed to and absorbed by ground. Weapon breaks down for backpacking, etc. 3) S & W suppressor/barrel assembly, supplied with vented barrel. Readily disassembled for cleaning. Baffle type with holding pressure chamber as in H & K MPS/SD design. 4) M16A1A suppressor. Superior to Sionics design due to internals that do not fail, larger capacity, 100% accuracy, and no relief valve which has proven to be counterproductive in sound meter tests as well as giving off significant flash in night firing. This unit is superior to all previous designs offered the public. Indefinite life with never a failure.

long life? We need to do more research in this area."

Exploring future developments, Ciener predicts, "The next form of suppressor development is the idea of sound transformation, in other words, raising or lowering the pitch to an inaudible level."

There have already been experimental designs in this approach to modifying sound, some as early as the 1940s. More recent experiments have been much more sophisticated, and some of the newer modes are currently under government study–and classification. For that reason it is difficult to present more definitive data on this method of sound suppression. But, Ciener is a man who will try to make his own data.

Born in Huntington, West Virginia in 1948, Ciener holds a degree in Business Administration from the University of Florida and served as a field artillery lieutenant in the U.S. Army. After his military obligation was completed, Ciener began full-time on the Grand Prix motorcycle roadracing tour. He recalls, "For every hour on the track, there are literally hundreds of preparation hours which require total attention to machine detail. This experience fine-tuned my design and machine operating talents."

After a severe crash in California ended his professional racing career, Ciener returned to Florida and managed a residential construction company. He also became a class III dealer in NFA weapons.

"This started as a moonlighting career. But I made so many contacts and sales, then started getting requests for custom silencer work, which was then unavailable, that I began to think about a new career," he recalls.

Ciener obtained the necessary licenses and approvals to begin his new full-time operation in the design, manufacture, and sale of silenced weapons.

"I'm now working at our new facility in Titusville, Florida, directly across the Indian River from the U.S. Space Shuttle launch facility. They make a lot of noise, but we don't," he adds with a touch of humorous irony.

Is Ciener successful? He says, "The hobby and part-time days are over We are almost always in a constant state of back order trying to fill the demand for my quality products."

Ciener also brings up a point which others often avoid when the question arises–cost. He says, "Customers always seem to ask me why the suppressor for their Mini 14, S & W M39 or M59, Walther PPK, or whatever, costs as much as or more than the firearm itself. Face it, suppressor manufacture is a highly specialized field which does not lend itself to mass production techniques due to the variety of weapons sent for suppression. Each weapon requires personal attention for custom design and installation.

"At present, the limited demand for suppressors does not support the expense of the special tooling necessary for mass production. However, as we standardize, there are some firearms for which we have sufficient demands for suppressors that some automated tooling is used, e.g. the Ruger Mark I, Colt M16, and the Military Armament guns. In these cases we can do a production run on semiautomated equipment."

Jonathan Arthur Ciener is willing to make these investments in the future of his field–suppressor design–because he plans to be around in his field for a long time. After all, as his ads say, "He's the best."

7. Do-It-Yourself Silencers

One theme of the 1970s was "do it yourself" and that handyman's anthem covered everything from holistic home building to surgery to intrapersonal religion to suppressor design and construction. In the realm of quieter firearms, entrepreneurship capitalized on the kit craze in 1979. The next step was to commercialize, then mass produce the easy-to-assemble kits. Soon all sorts of do-it-by-the-numbers suppressor products appeared.

Since 1980, about a dozen small companies have been formed to sell suppressor components and kits. Advertising heavily in *Shotgun News* and operating mostly by direct mail order, they offer all the necessary parts to build a suppressor. Interestingly, their ads have gotten personally competitive, too, and not only in price and quality. In 1982, one company was running ads that took personal swipes at a competitor's products and ethics, threatening legal action over alleged business intrusions. The other partly claimed he was a mirror-image victim of the first.

Many of the kit companies opened for only an ad or two, then quietly folded back into the graveyard of failed small businesses, a sad legacy of our times. Others reorganized, renamed, and carried on. Still others were mail order ripoff artists who grabbed the bucks and ran. Most remained Ma and Pa moonlight operations. Yet, a few that opened for business at the kitchen table with a post office box became actual companies when they combined with a machine shop subcontractor. Some of them have made good money.

Basically, these companies sell only component parts. Most solemnly advise in each ad that actually assembling or "activating" a suppressor requires government approval and tax payment. All send warning notices with their kits; some even furnish the necessary government forms to be filled out when the buyer applies to "activate" the device.

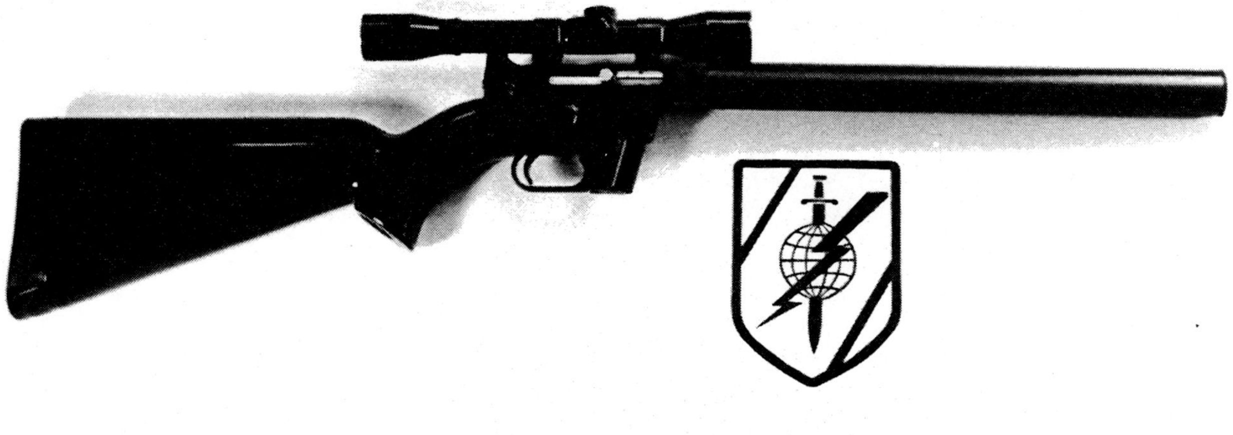

SPECIAL MISSION GROUP

One of the most suppressor-mated weapons—the AR7 using an integral SMG suppressor.

But, few folks are fooled by this veneer of "we-just-sell-'em, we-don't-know-what-the-customers-do-with-'em" logic.

The Bureau of Alcohol, Tobacco and Firearms eventually joined the game. My attempts to get official or unofficial reaction to the kits from ATF were not very rewarding until February of 1982, when Edward M. Owen, Jr., chief of their Firearms Technology Branch, offered the following explanation of the legal status of the various "silencer kits" being offered for sale:

> Due to the varying designs and construction of the kits currently offered for sale, we can only render an opinion based on a physical examination and testing of the specific kits. Those kits which require no additional manufacturing to assemble a functional silencer would obviously be subject to the National Firearms Act (NFA) and those which require additional manufacturing steps can only be classified on a case-by-case basis. The factors which must be taken into account are complexity of operations, tools, and the time required to produce a functional silencer.

However, any reasonably mediocre advertising writer could easily weasel around those ATF semantics, all in the spirit of American can-do capitalism, of course. For example, in 1981 an outfit known as R.F.P., operating out of Milpitas, California, offered an Uzi suppressor kit for sale at $150, along with the warning, "Sold for repair or display only." The unit was almost a duplicate of the original Sionics/MAC suppressor. They also offered kits for the MAC 10 and 11, Ruger pistols, rifles, and the AR7 at $100 each.

R.F.P.'s president, Robert F. Pierce, most recently designed a new wipe system called Encap-U-Wipe. Patent is pending. He says that this attachment will drop the dB rating another three points. All of their units, except those in .22 caliber, carry this new device. Pierce claims, "Because of the added third wipe, our kit currently has the lowest decibel rating of any sonic type kits on the market."

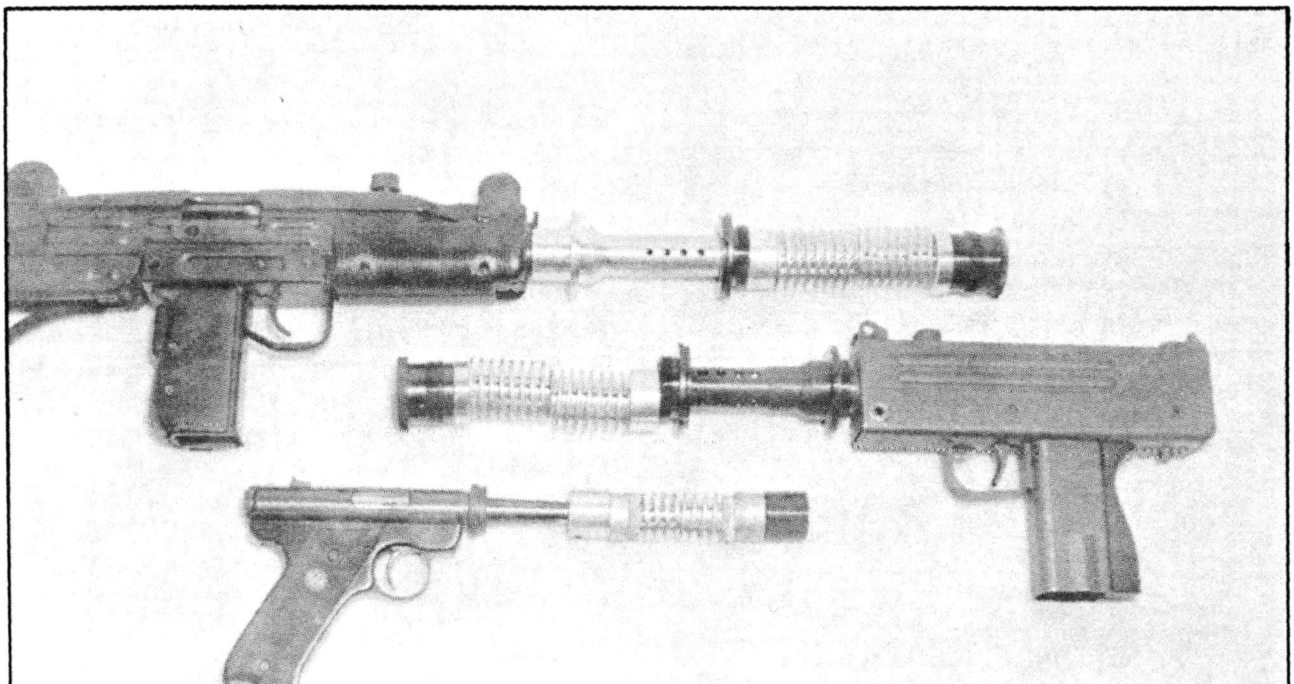

R. F. P.

R. F. P.'s Uzi kit, the MAC 10 kit, and the Ruger RST 4 kit. Their Uzi design is unique in that the suppressor kit slides over the short barrel, replacing the barrel nut, which keeps the overall length to a *minimum*.

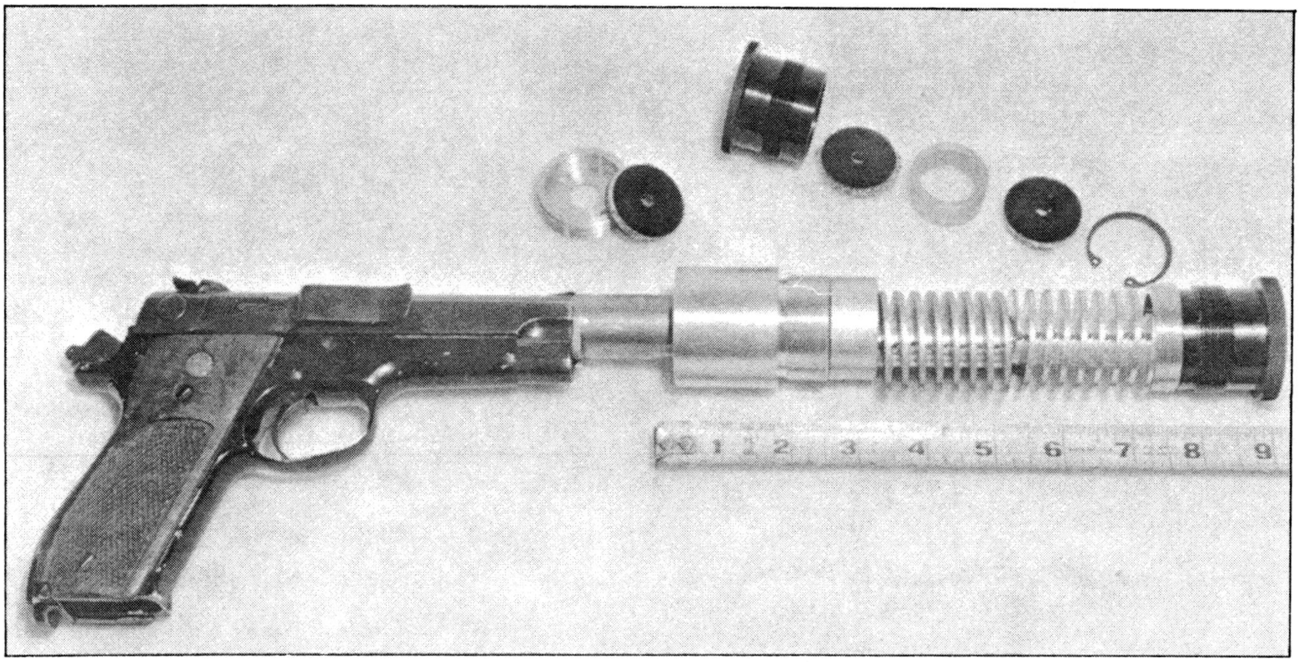

R. F. P.

The R. F. P. mini kit installed on a S & W 39. This unit carries R. F. P.'s new Encap-U-Wipe, a new wipe design on all their sonic-designed suppressors.

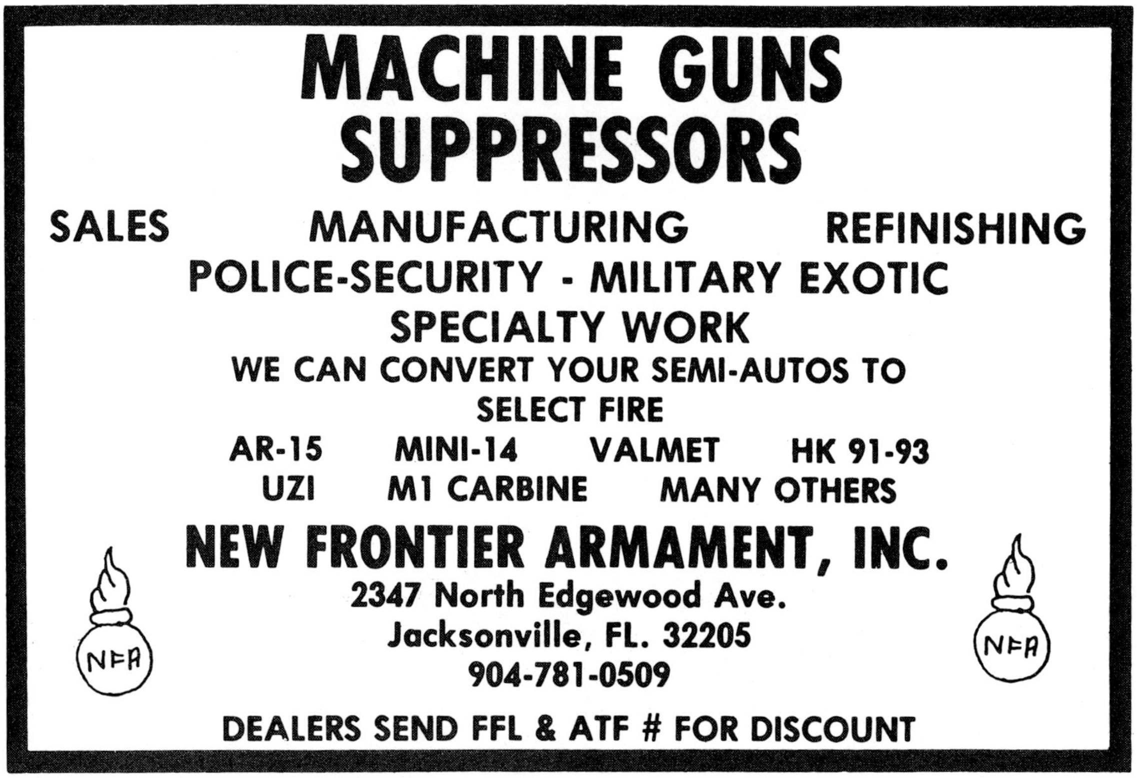

SHOTGUN NEWS/NFA

One of the cottage industries. An independent doing custom work, including suppressors.

A. Z. SANTINI

Colt Woodsman with NS stock.

A. Z. SANTINI

Llama with long bushing attached to barrel to allow suppressor to be attached.

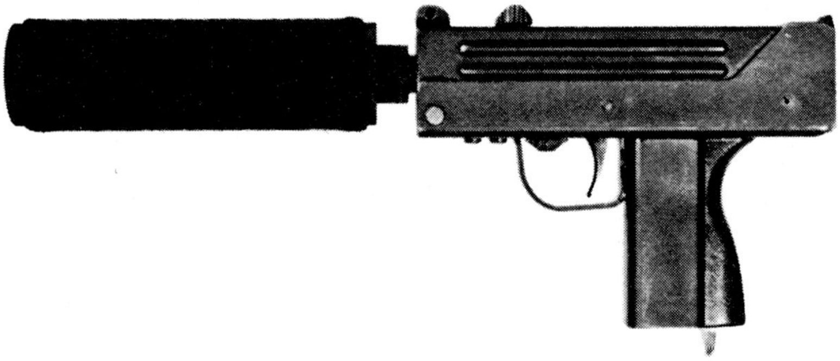

A. Z. SANTINI

.45 MAC. Note the length in comparison to the original.

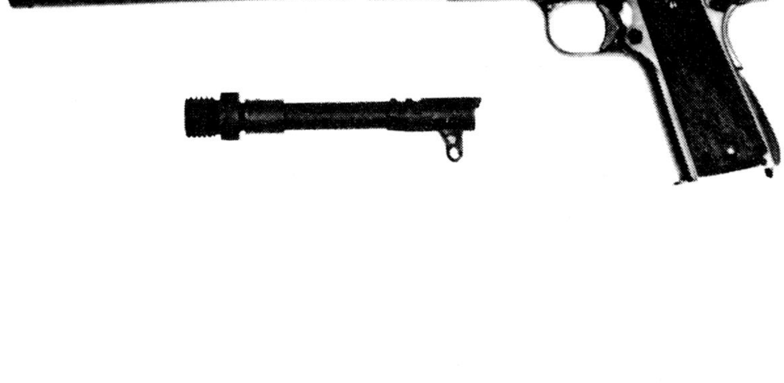

A. Z. SANTINI

Colt .45 shown with long barrel made to adapt the MAC suppressor to it.

R. E. CHOATE

R. E. Choate with his MC-1 suppressor on the Supermatic. He is also holding the internal guts of an MC-1 unit.

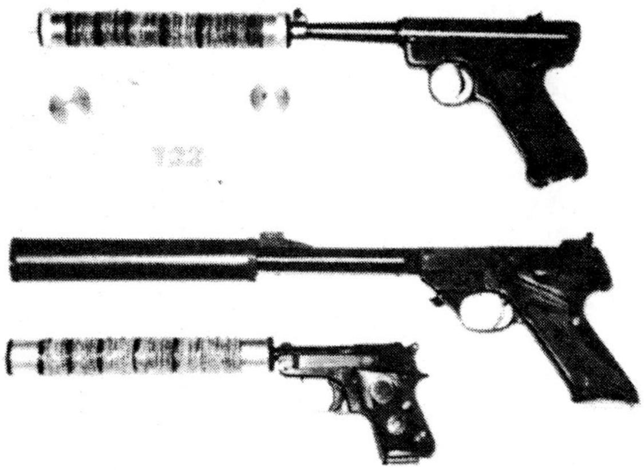

R. E. CHOATE

Three .22 caliber handguns with the Choate-designed muzzle suppressors. Top to bottom: Ruger target pistol, High Standard Supermatic; Beretta Minx. The suppressor is shown with and without the outer tube and can be used on any .22 with a properly threaded barrel.

R. E. Choate is a cheerful, friendly man who is an expert machinist, custom gunsmith, and according to a business competitor, "a first-class guy who is a mechanical genius." In the suppressor field, Choate designs, manufactures, and sells kits. He also does barrel threading, custom mill and lathe work, plus, as he says, "I'll try anything I can to help anyone."

He sells his MC-1, .22 caliber suppressor kit for $120. The assembly kit goes for $8. He also sells caps, screenings, discs, and wipes for his and other suppressors, including the MAC, Uzi, KG 9, AR15 and M16, AR180, and, believe it or not, the M60.

He has also designed the MC-2 kit in both .380 and 9mm. His most recent custom design was for an AR180, which is so efficient, he claims, that cleaning and screen replacement need only be made after 3,000 rounds or more. His designs are clean, efficient, and well thought out. His machine work is top quality and the reputation of the Choate units is excellent, according to users and others who've tested them.

Survival Supply Company is a Class II manufacturer of suppressors that turns out custom designs for "almost any closed breech weapon," according to owner Gary Wilham. His designs utilize steel tubes with aluminum end caps using wipes. Screen wire, foam, and stainless steel scouring pad material make up the internal parts of the SSC suppressors.

"For example, my AR7 suppressor consists of a ported barrel, a rear end cap, a steel outer tube, and a front end wipe assembly. The wipes are replaceable, of course," Wilham says. "The internal workings consist of screen wire over the first two sets of parts in the ported barrel area, then oil-soaked foam rubber for the rest of the length of the tube. The unit is approximately 65 to 70 dB.

"With high power units I use a bit different technology, but I do not go to the old Sionics' static-type spiral system. My screening soaks up the most violent part of the expanding gases, then the scouring pads act as a dynamic buffer."

Wilham makes one six-inch suppressor for the AR15 which can be fired easily without ear protection and heard only as a normal .22. He adds, "When I increase the length to twelve inches, the unit quiets the weapon so much that all you hear downrange is the ballistic crack of the round."

He makes a variety of custom designs including some for shotguns. Wilham says, "I have one suppressor for a single-shot H&R 20 gauge that makes the weapon sound like a capgun."

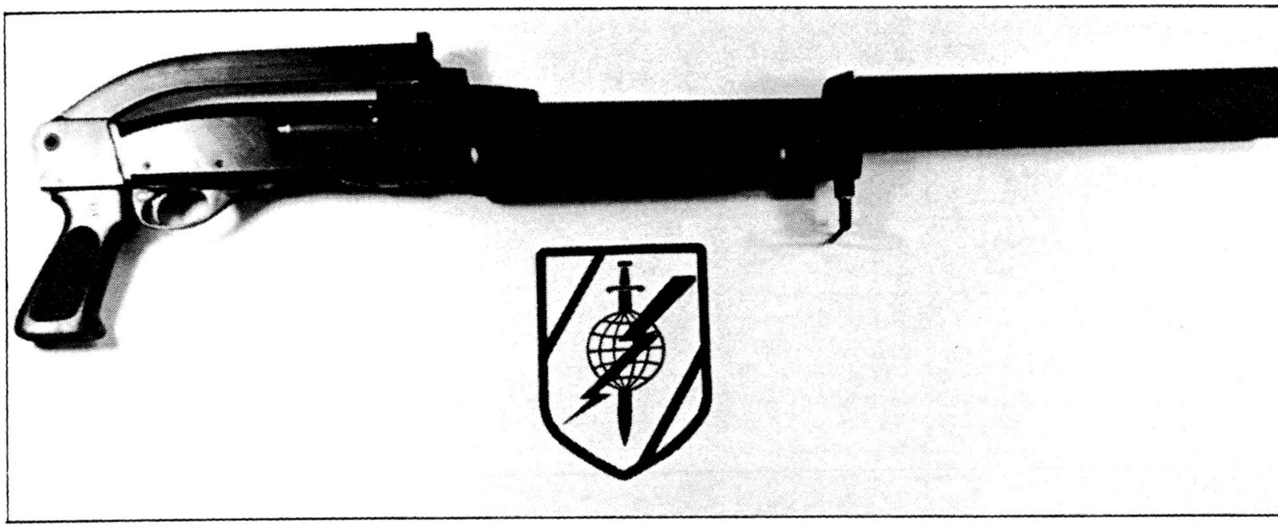

SPECIAL MISSION GROUP

Here's a rarity . . . a suppressor-equipped shotgun. In this case, it's a 12 gauge Remington 870 using a SMG suppressor.

As Georgia seems to be the leading state for pioneer suppressor inventors, designers, and manufacturers, it's not surprising that Mitch WerBell, Sionics, Military Armament Corporation, et al. all had their roots in that fine state. Perusing today's silencer parts and kits ads shows that many of these current businesses are homegrown there, too.

For example, Brown and Associates from Cumming, Georgia, sell what they term "the finest quality [tubes] available today." Theirs are replacement outer tubes for the Ingram, Sionics, and MAC submachine guns. They also sell tubes for a variety of other weapons, but do not market any internal parts. One can buy those from some other company. Tube prices range from thirty dollars for a Colt Woodsman to sixty dollars for the AR15, M14, HK91, and HK93 units.

In 1981, an outfit known as Catawba was operating out of the old WerBell/MAC home country in Marietta, Georgia offering Sionics replacement tubes. In full page *Shotgun News* advertisements, they pledged "The world's leading tube manufacturer offers you the very best at tremendous savings." By late 1982, they had opened a retail showroom and store offering a variety of suppressor tubes ranging from a Ruger and High Standard pistol model at twenty-three dollars to the AR7 model at forty-seven dollars. Two columns listing tubes for various weapons covered the middle of their large ads—some firearms I had never seen with suppressors before.

They also noted, "The Bureau of Alcohol, Tobacco and Firearms has determined that the above items alone are legal. However, these items in conjunction with certain other parts may be subject to the provisions of the Gun Control Act of 1968 or the National Firearms Act."

Catawba also offered the following security to its customers: "We are not required to hold any firearms licenses and do not hold any. We are, therefore, keepers of confidential records and are not subject to record checks." That means the ATF could not seize Catawba's customer list and go on a witch hunt to find illegal suppressors.

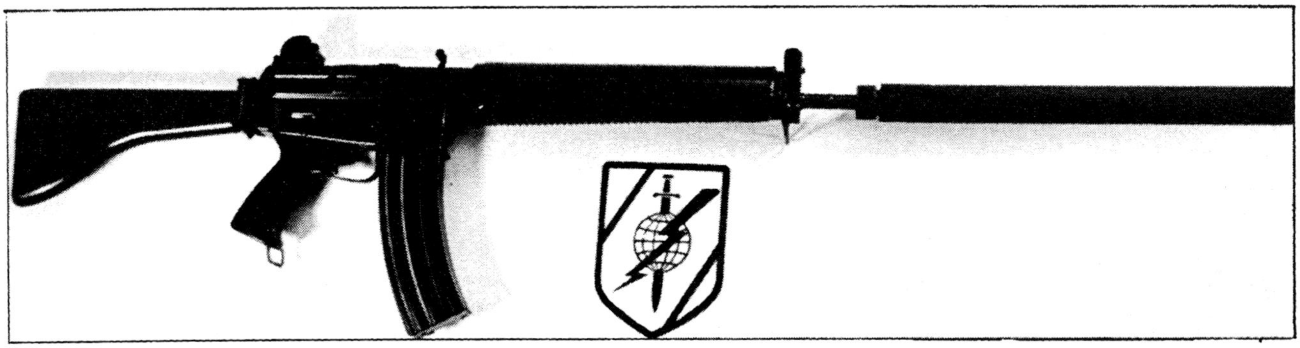

SPECIAL MISSION GROUP

The Armalite AR 180 carrying a completed .223 suppressor from SMG. The same unit will fit the AR15.

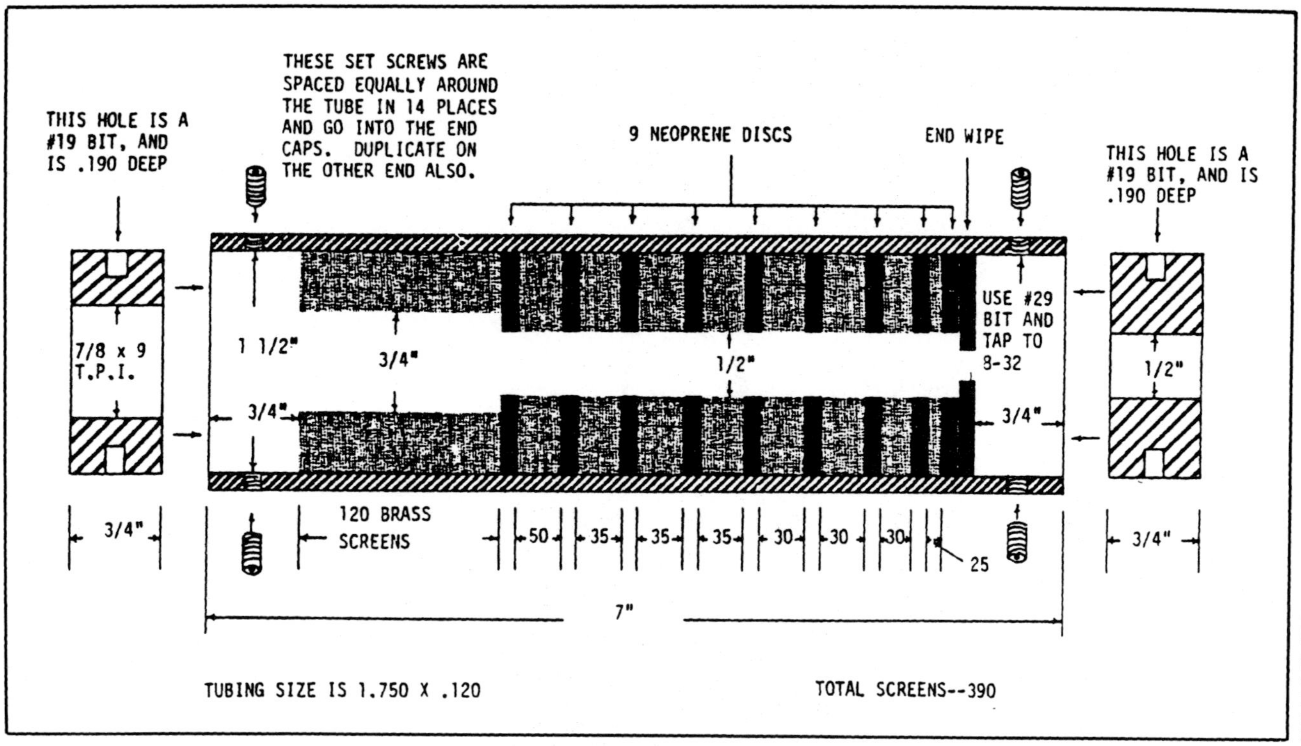

A. Z. SANTINI

Diagram of the internal design and construction of the Sionac suppressor in .380 and .45.

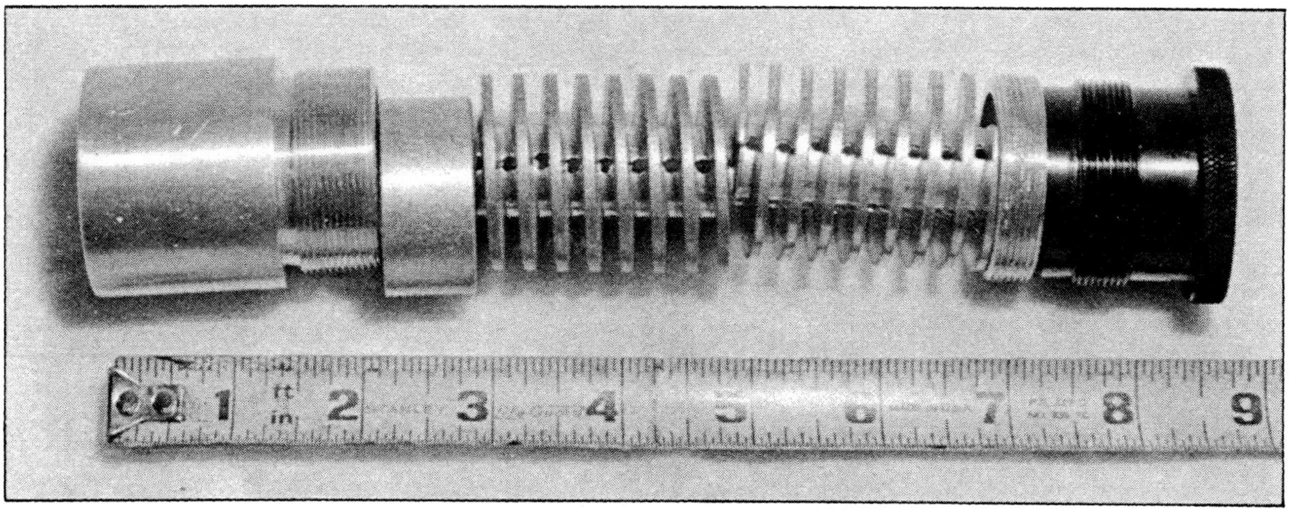

R. F. P.

This is the standard R. F. P. mini kit using a MAC 10 outer tube. The unit is about nine inches long and will fit various weapons.

LEE THOMAS, MAC

The "by the numbers" quality kit work by Military Accessories Company (MAC) is, in this case, the guts of their suppressor for the Ruger 10/22.

By late in 1982, Catawba had expanded its offerings into other suppressor paraphernalia, plus exotic weaponry parts and machine services. They still are not a firearms dealership and still hold no licenses, much to the relief of their customers.

One businessman, however, Don Floyd of Woodstock, Georgia, offered complete Sionics-type suppressor kits for sale in his 1981 advertising. Prices ranged from $140 for MAC M10 and 11 units to $175 for the M14, AR15, and HK93 suppressors.

Another of the major manufacturers who "WerBellized" a trade name is Lee Thomas of Military Accessories Company (MAC) of Cleveland, South Carolina, who owns a full-line production business dealing in suppressor components, accessories, and publications. MAC's work is all top of the line with quality machining and finish. In 1980, operating from Atlanta, their ads opened with parts for the Ingram weapons and suppressors, plus suppressor components for the Charter Arms survival rifles. One of their suppressor ads noted, "If you ever need to rely on your AR7 in a future survival situation, there may be times when you will not want others to know your firing location."

In addition, the ad, which offered the AR7 muzzle adapter and internal suppressor parts for $175, carefully noted that prior ATF approval was needed to own and assemble a complete unit. MAC also said, "Since we hold no government firearms license, we are not subject to any form of record check, as are other dealers, and all our sales are kept confidential."

The following spring, before MAC moved to South Carolina, they offered *Shotgun News* readers grand values during a silencer "clearance-moving" sale. They also offered a no-frills, easy-to-understand warranty on their products of "full refund or replacement, no questions asked." Thomas added, "We have total faith in our products, so we can easily live with that warranty." They also offered original Sionics sniper rifle suppressor parts in both .223 and .308 for $200 per set, "a real vaue for user or collector."

By that summer, MAC's quality product line included adapters and kits for the standard Colt, High Standard, and Ruger pistols that went well with suppressors. Prices were competitive. Their suppressor internal parts were all made of precision-machined barrel steel, while the barrel bushings were made of brass to protect the gun's bluing. In tests the MAC units performed well, showing them top quality suppressors.

LEE THOMAS, MAC

Note the WerBell influence in this counterswirl design—the Military Accessories Company's internal suppressor parts for the Ingram M10.

LEE THOMAS, MAC

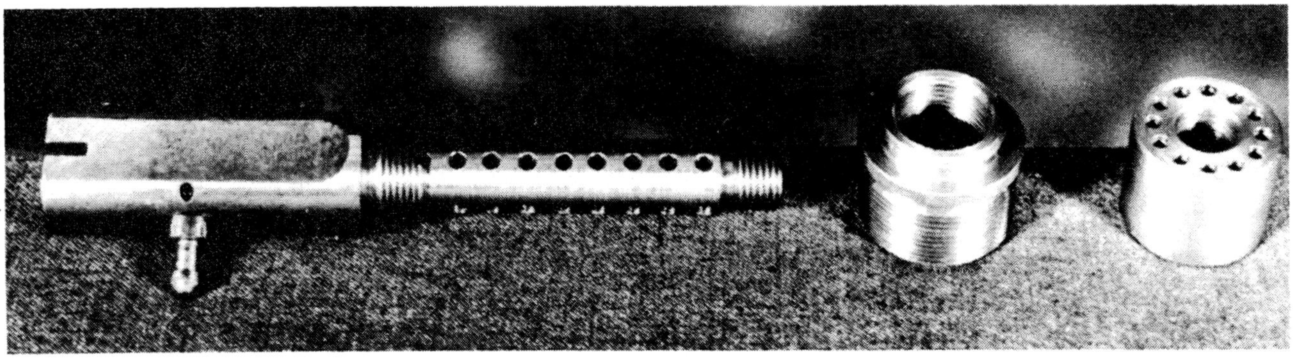

This finely machined piece of equipment is the MAC internal parts kit for a .22 High Standard pistol.

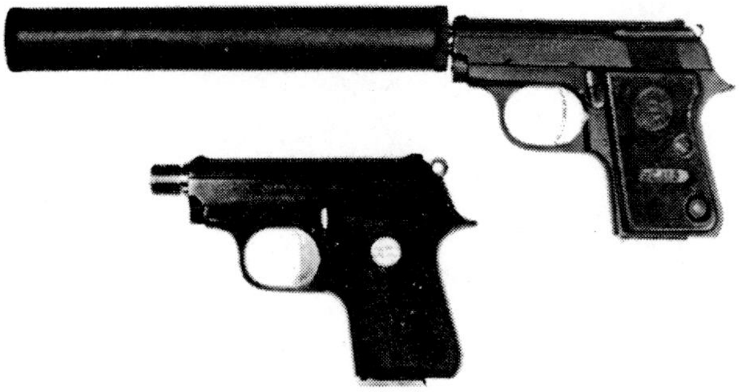

A. Z. SANTINI

Astra Cubs with new barrel showing.

A. Z. SANTINI

Beretta .22 with extra length bushing and thread. The end cap is a thread hider which can be screwed onto the thread when not using suppressor

SIONAC

Sionac technician with a Sionac suppressor on a MAC M10 .45. The suppressor is half the length of the original suppressor but much quieter, according to Santini.

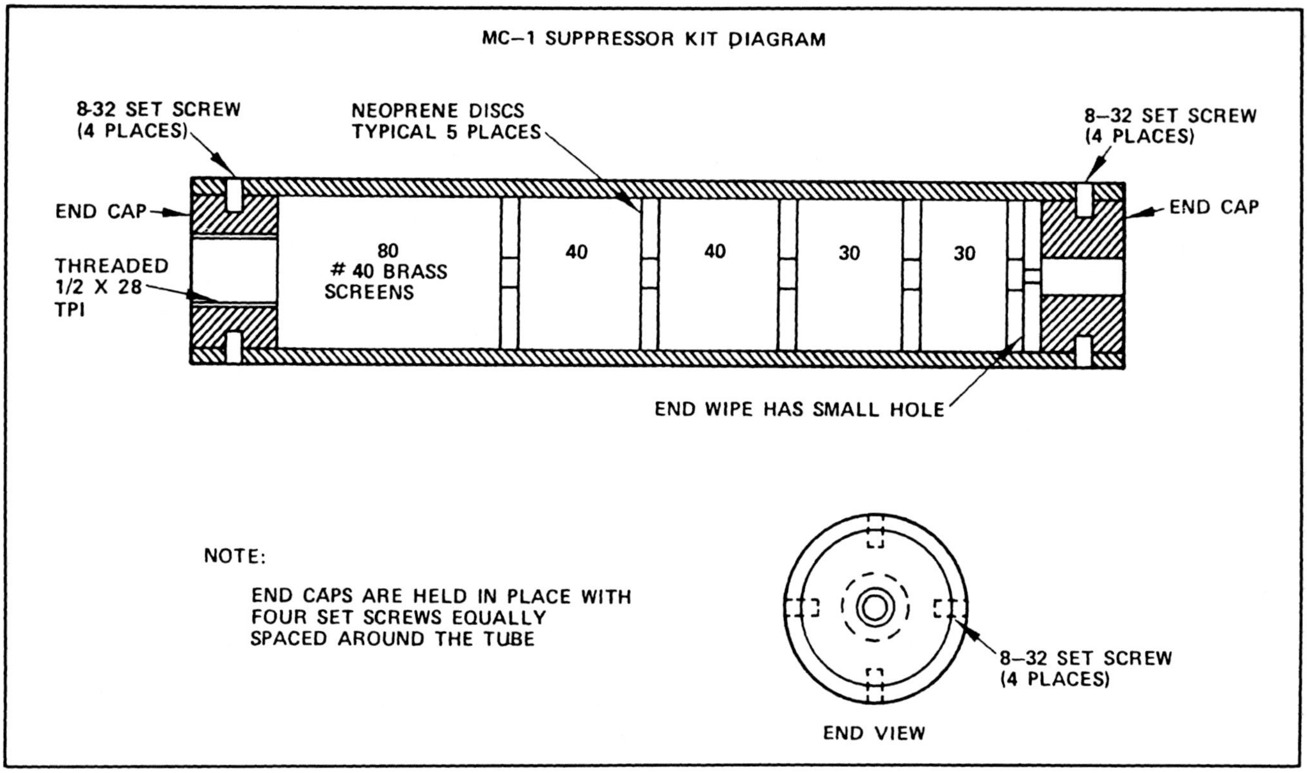

R. E. CHOATE

Engineering diagram of the Choate MC-1 suppressor kit.

In May of 1982, MAC expanded their product line to include a variety of clamp-on muzzle adapters for various pistols, as well as conversion adapters to allow the use of M10 or M11 suppressors on most .22 caliber pistols. They also offered custom ported High Standard barrels. By September they had completed designs of a newly modified High Standard model which they advertised as, "When completed, [these units] are without question the most effective suppressors on today's commercial market. Satisfaction is guaranteed or your complete purchase price will be refunded."

Prices for the internal component entire package, less the outer tube, ran from $260 for a 5.5 inch to $325 for a 7.25 inch barrel with custom mount and 3X scope.
scope.

Another kit for the various High Standard .22 caliber pistols was offered late in 1982 by J. R. Stemple, who advertised all internal parts and wipes, including a ported barrel, for $89.50. He also offered an extra long .45 M1911 barrel with bushing and integral muzzle coupling for an M10 suppressor for $77.50. Mr. Stemple advertised various other couplings for such weapons as the Uzi, Thompson Contender and Ruger handguns. He also offered a variety of random tubes suitable for the M10, M11, M1, M14, HK 91 and 93, Colt Woodsman, and others with prices ranging from $19.50 to $36.50.

Other dealers offer accessories only. For example, Shooters Equipment Company of Tamassee, South Carolina, sells a splendid replacement for the original Sionics/MAC wipes, which are notoriously inefficient. The new Accu-Wipe costs twenty dollars, is indestructible, needs no replacement, and is easily installed.

Not all suppressor news comes from Georgia, however. Sionac, a quality-conscious company in Tucson, Arizona, first used a modest classified ad to offer *Shotgun News* readers "the best .22 LR unit in the world" for $125. This included all the interior components, but not the outer tube. According to ATF interpretations, the numbered outer tube is the legally registered suppressor.

Working with sophisticated customers in mind, Sionac offers the booklet *The Silencer: A Report Detailing Construction of an Effective .22 Caliber Suppressor* for eleven dollars. The Sionic ad reads, "Easy-to-follow plans not requiring machine tools outline construction of a suppressed weapon measuring 72 dBs." If you order their parts pack, you get their free, excellent, well-written no-nonsense publication.

By the end of 1982, Sionac claimed that mass sales and production allowed a price reduction, so they cut the price of their parts pack to ninety-five dollars. Their ads claimed a sound level comparable to a BB gun and strongly pitched their quality and technology. One point made a lot of sense. Sionac notes, "Don't get fooled into purchasing junk-in-the-box. Write us a letter or give us a call and ask all the questions you want. We have the answers." It's a valid suggestion and a good point.

Part of the research for this book included contact with law enforcement people who had purchased various kit units for evaluation purposes. I saw and tried many of their units. Most ranged from awful and beyond; some were fairly effective. Most tended to shootout after four hundred or five hundred rounds. In some the barrel alignment was dangerously off. One ported unit lodged a round in the barrel. For the most part, these mail-order suppressor kits were more monuments to avarice than ballistic sound suppression technology.

This is notably not the case with the Sionac units. They are professionally done, top-quality products. Their performance in suppressing sound exceeds some of the commercial designs being tested for military and clandestine use by the U.S. government.

According to A. Z. Santini of Sionac, their .22 suppressor is seven inches long and an inch in diameter. It is a baffle type unit made of aluminum, consisting of nine chambers with eighty mesh brass screen washers, ten neoprene baffles, a special

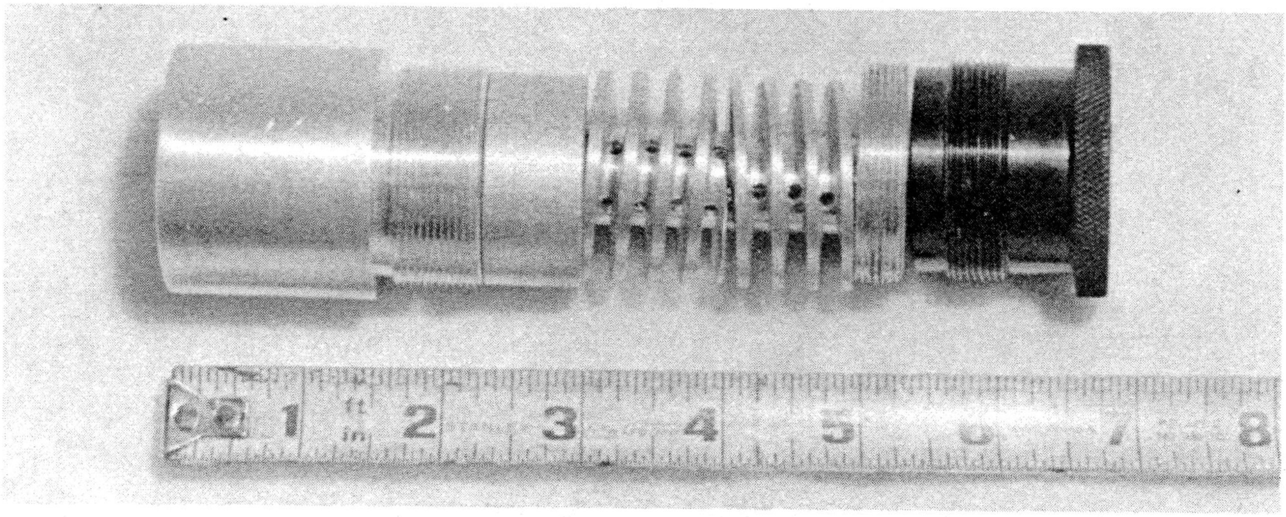

R.F.P.

The R. F. P. super mini kit for the .45 M1911A1 or the S & W models 39 and 59.

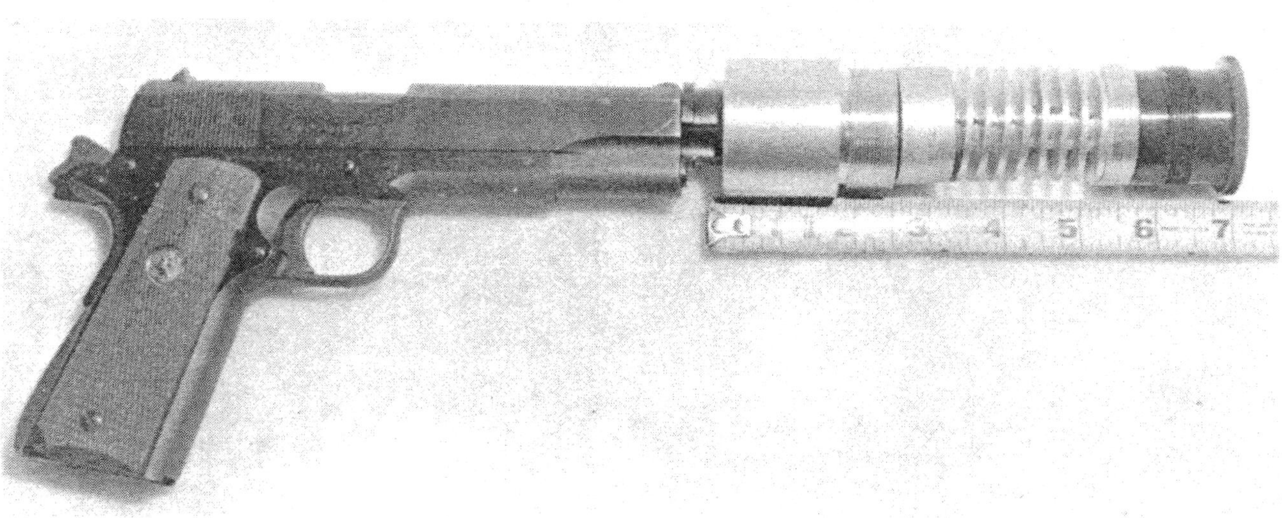

R.F.P.

R. F. P.'s super mini kit and adapter bushing installed on a .45 government model.

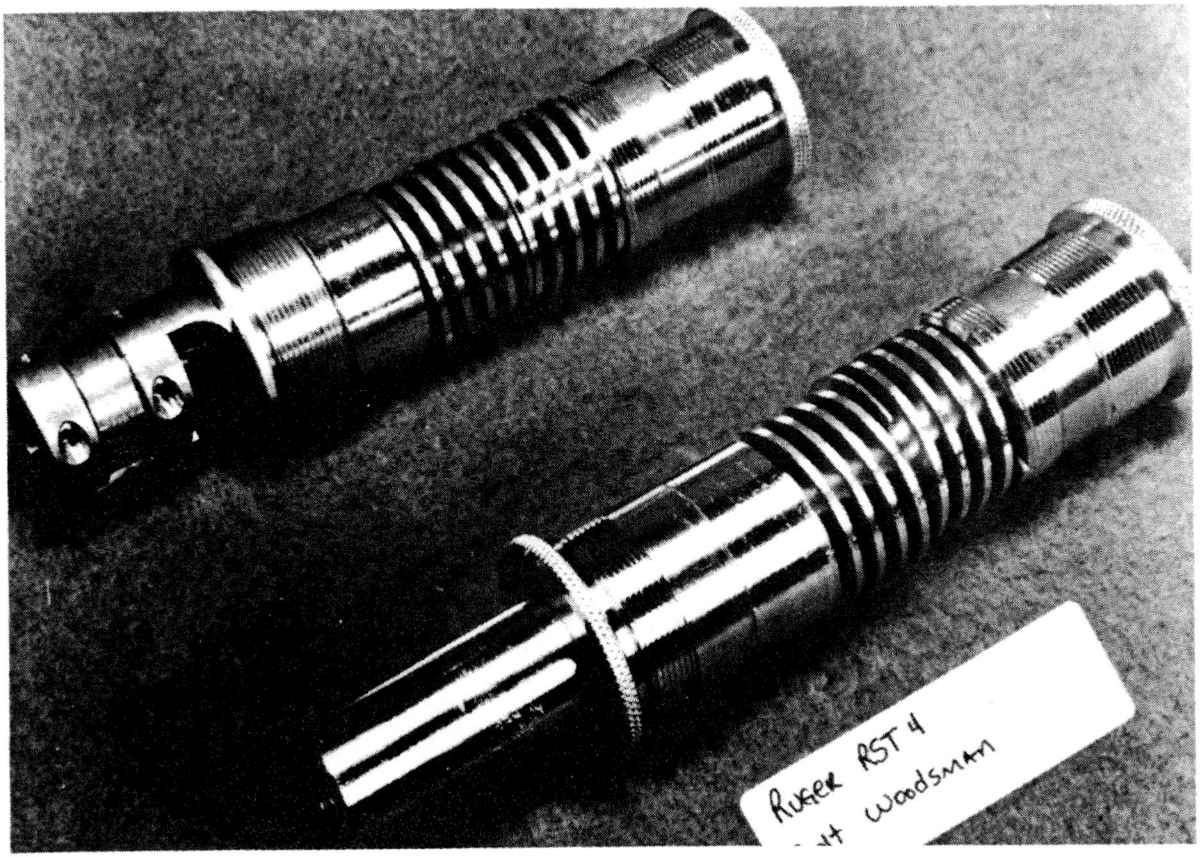

LEE THOMAS, MAC

The MAC, high quality suppressor kit for a Ruger .22 RST 4 (top) and Colt .22 Woodsman (bottom). All work on these units is top quality.

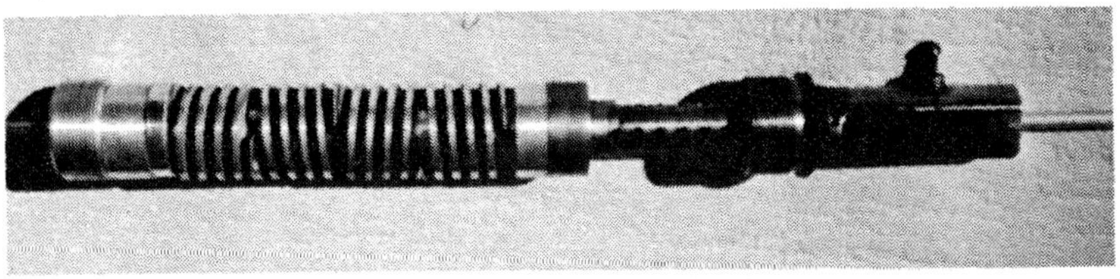

E. HOWARD COLBY

Government agencies buy and test suppressor kits in hopes of getting sanitized equipment. This MAC PAC kit (left) tested out "very quiet with subsonic ammo . . . loaded to 675 FPS only primer firing is heard," according to one CIA report.

HD military model (right) with Sionac parts pack modified for use with drilled and ported barrel. According to Santini, the dB level is louder than with other styles of moderators, but accuracy is improved.

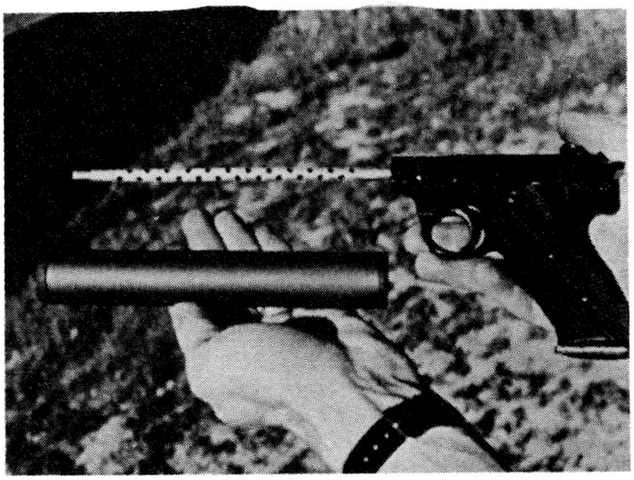

SIONAC

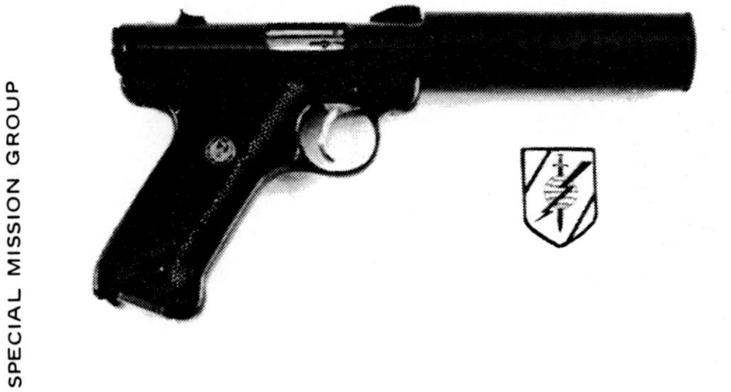

The completed SMG suppressor kit on a Ruger RST-6.

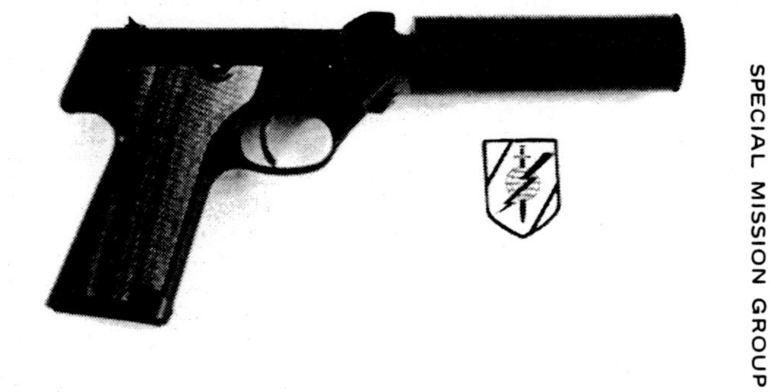

A silenced High Standard Sport King with the completed SMG suppressor.

end wipe, plus the end caps. Sionac also offers a suppressor parts pack for the .45, 9mm and .380, obviously designed for the Ingram weapons. In 1982 they added a threaded, long barrel for the M1911A1 so it could be used with an Ingram style suppressor.

In terms of testing, the Sionac .22 caliber suppressor indicated a dB level of 70, while their larger .380 and .45 units came in at 75 dB, which is about the level of electric typewriter noise. For comparison purposes, Sionac's booklet describes as "painful" the 130 dB of a pneumatic hammer, while the threshold of human hearing is 5 dB. Of the kit suppressors I have seen, the Sionac models are among the best, the rival of commercial designs.

The Special Missions Group, Ltd. first appeared in print in January of 1981 with a *Shotgun News* ad selling suppressor kits for the Ruger RST and the High Standard pistol for $125, plus the AR7 and a universal .22 rifle model for $150. All parts were 4130 ordnance steel or T-6 aircraft quality aluminum. Listed as a post office box in Soda Springs, California, the Special Missions Group offered full instructions, including those for legal activation. An inquiry to the company brought a cryptic response initialed "P. D. S." from one who called himself general manager. "Company policy is not to release personal information or particulars about the company to journalists or members of the press," he wrote.

Later, however, D. W. Lambert, Special Mission Group product design director, did note that the company had a prototype suppressor for the AR15/M16 and the AR180, claiming sound reduction to that of a .22 rimfire. P. D. S. noted that their Ingram suppressor was "very slightly more effective than the factory unit." They were also developing a suppressor for the Remington 870 shotgun, but no details were available at that time. But by May of 1981, Special Mission Group was advertising its Remington 870 suppressor kit for $225. A Mini 14 unit was also going for $225, as was the AR15/M16 suppressor kit. Their Ingram M10 model sold for $180. All parts were included and each unit had a Parkerized finish.

In their activation instructions, the Special Missions Group people noted at the end of the operational steps, "Your suppressor is now installed in an inactive condition (Hollywood style) and will not suppress your weapon. Many of our customers prefer this as it avoids many problems. They wait until they are outside of U.S. territory to activate their suppressor."

The most flamboyant and humorous kit designer and seller is Detroit's George Dodson, who bills himself as "The Invisible Man." Anybody who has seen his attention-gathering "SU PRESS ON" ads will not soon forget this businessman. Claiming to be the man who invented the autosear for the AR15 and AR180, Dodson has a suppressor that "snaps and locks" on the barrel of the Ruger pistol. It's a good idea, as there is no need to drill or thread the pistol; just slip Dodson's unit onto the barrel and it locks in place until you want to remove it. He sells his Ruger units for ninety-five dollars. They are tough, of solid construction, and handle well when activated.

A set of instructions for assembly, plus the obligatory BATF warning, is included with each component parts kit. Dodson says all of his units can be assembled with hand tools, making the home workshop effort easier than with some kits. Dodson's instructions are clear, concise, and amusing to the point of black humor about the consequences of being nailed by the ATF for producing an illegal suppressor.

In addition to the sales rush for 1980s technology and the claims race for more and better silence, history has not been overlooked. Nostalgia aficionados no doubt got misty-eyed at the *Shotgun News* ad of D.A.Q. in Cicero, Illinois, offering a complete rebuild kit for those .22 caliber Maxim Model silencers that were not the sealed units. Or consider the excitement of collectors over the 1980 ad of G.V. Metal Products in British Columbia, in which a cutaway version of the rare Maxim Model 15 silencer was offered for sale at eighty-nine dollars.

Another brand of business also opens when the real thing is restricted by zealous government authoritarians. Lookalike and replica silencer products came on the market shortly after the first real kits became popular items.

Replicas have been around the firearms business for years, ever since a fine gentleman named Tom Nelson pioneered the concept in this country shortly after the 1968 GCA. Replica suppressors were first advertised in 1979, when the Garfield Target Range in Garfield, New Jersey, offered a suppressor lookalike for the MAC model submachine gun at $31.50. The following year, Security Programs, Inc. across the river in Philadelphia, Pennsylvania, sold much the same thing at $39.95. Neither item would perform the real function of a suppressor, so there were absolutely no restrictions on their sale.

The home workshop market seems to be a strong, profitable one. The companies that prosper in this field have the same characteristics as those that do well in any field: they offer a quality product at a fair price and deal honestly with their customers. Some of the silencer kits are ripoffs; some designs are no better than a competent person could do with his own hand tools and some steel. Yet some kit makers approach the work being done by the commercial people both in terms of workmanship and effectiveness; a few designs and units surpass that.

Although no official will confirm this, several insiders have told me that the government has bought and tested a number of the kit designs and that some are under active consideration for limited supply orders for special mission purposes. However, for the most part, these kits are being bought by tinkerers, experimenters, and people who just want to see what a silencer is and how it operates. Curiosity, rather than malevolence, seems to be the major factor behind most of the purchases. One ATF supervisor I spoke with said, "Sure some of the bad guys will try some of these mail order kits, but the vast majority of buyers are the same types of guys you're going to see at a gun show. I'm not too worried about them at all."

Isn't that a refreshing attitude? It's too bad it's not the official one.

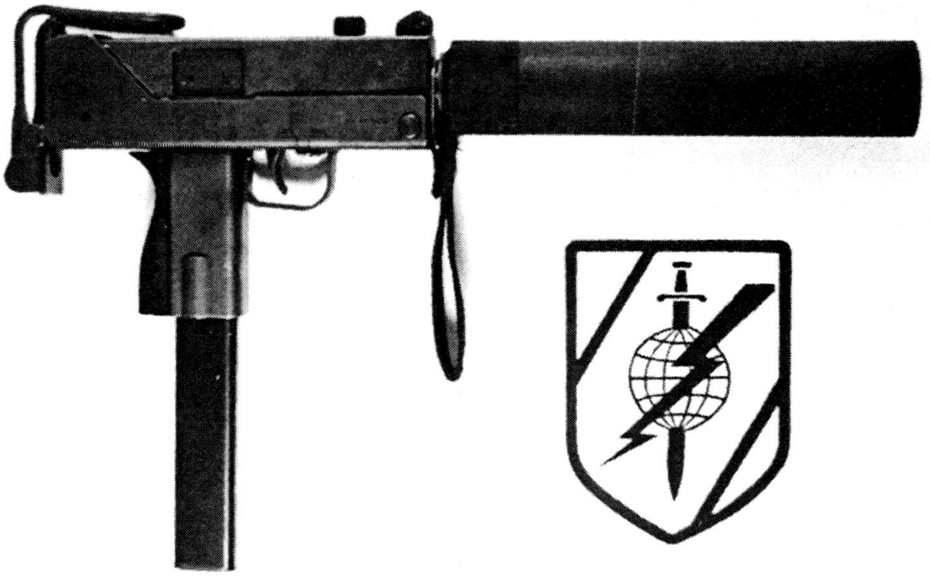

SPECIAL MISSION GROUP

The Ingram M10 with SMG suppressor made from their kit.

SPECIAL MISSION GROUP

The Mini-14 with SMG suppressor is reportedly a very efficient unit.

J. DAVID TRUBY

The "Invisible Man's" slide-on suppressor kit.

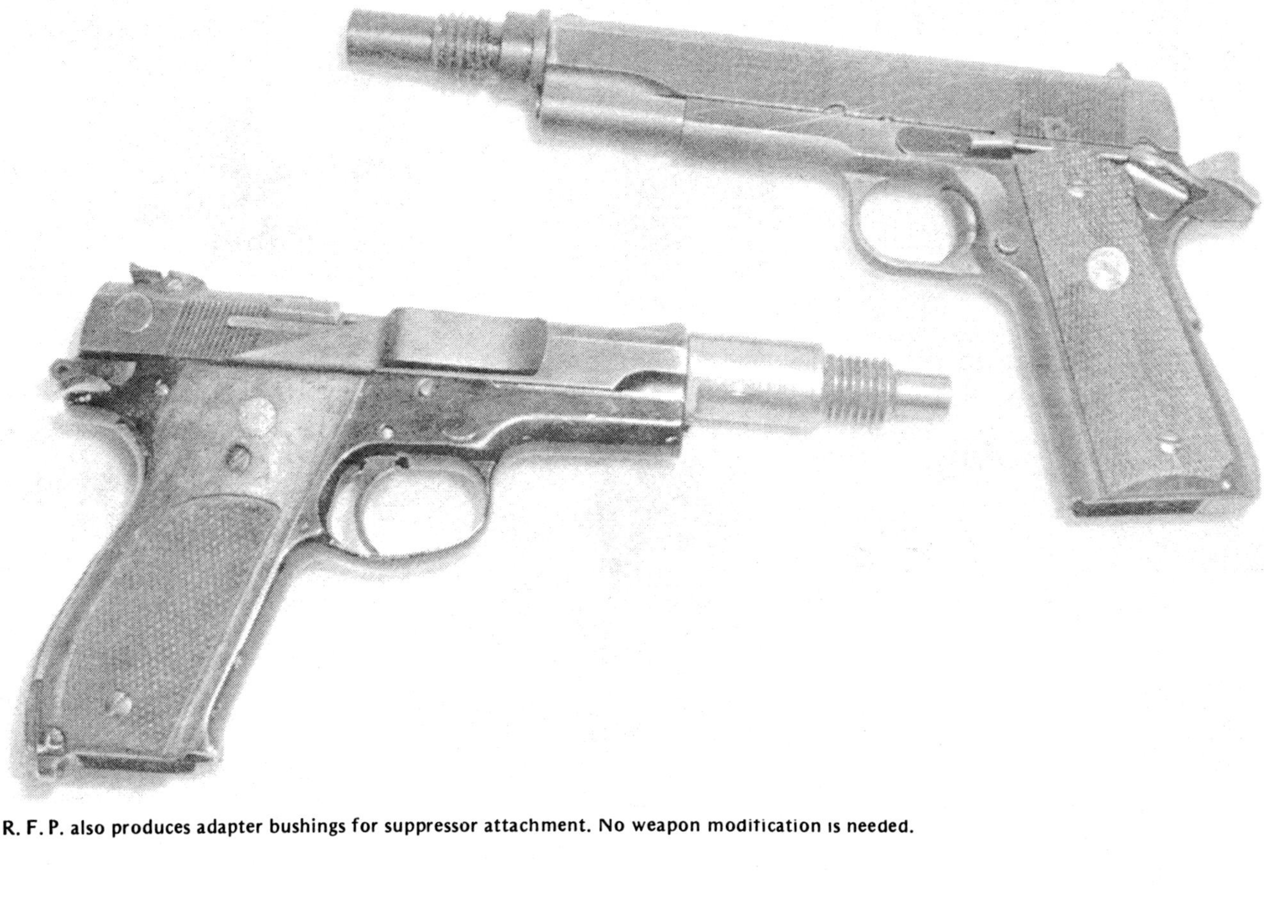

R. F. P.

R. F. P. also produces adapter bushings for suppressor attachment. No weapon modification is needed.

8. Boobs, Tubes, and Dollars Wallow in Politics

In the early 1970s, an apparently feckless-seeming group of planners in the Pentagon started marching out of cadence with the normal lockstep of ordnance procurement. The accepted method was still the old boy network of cronyism, featuring active duty officers, recently retired officers just hired by ordnance sellers, plus the other dollared trappings of what Col. Isaac Lewis once termed "The Ordnance Ring." The planners' new idea was to consolidate, coordinate, and cut cost. In 1977, this new method was initiated amid a shower of publicity. It was known as the Joint Services Small Arms Program (JSSAP).

According to a JSSAP spokesperson, the purpose was to "consolidate all small arms research, development, testing, and evaluation, including suppressor requirements, into one program for all five armed services." This was supposed to reduce horrendous research and development costs, eliminate duplication of effort, and drastically cut the lead time between weapon concept and adoption. JSSAP was also designed to cut the usual and wasteful interservice rivalries, the empire building and turf battles, plus the Ordnance Ring buddyism between military procurement leadership and the arms industry.

"JSSAP was going to get better weapons into the hands of the troops faster, at less cost, and more efficiently," the spokesperson explained.

One of the first activities was to announce a series of open trials for an entire family of weapons, ranging from pistols to heavy machine guns. JSSAP supporters said that the trials would be totally open and free of the old Ordnance Ring skulduggery of the past. In JSSAP's brief life, as this book is written, the veracity of that claim remains murky, according to most sources. According to a 1982 study by the General Accounting Office, noncompetitive methods were still the basis for more than 50 percent of Pentagon contracts, totalling more than $80 billion in 1980. In 1981, the figure was nearly 70 percent.

But when the program started, ordnance observers felt that where there was JSSAP, there was hope. Some of the respected experts thought that the military should get back to basics and away from the expensive and unreliable weaponry of the immediate past. Col. David Hackworth, the U.S. Army's most decorated soldier during the Vietnam war, stated, "Unless cooler heads prevail, during the next five years America will spend $1.5 trillion [that's trillion] replacing Vietnam era junk with a new generation of junk. . . . In Vietnam, we painfully discovered that all of our bombs, bullets, and miracles could not get us into the winner's circle. We also discovered that we had spent billions of dollars on crash programs that produced highly sophisticated junk."

Sid Taylor, a former U.S. Air Force systems management expert, said, "Unless we rethink, reconceive, and simplify some of our goldplated, overly complex, overly costly . . . military weapons systems, the fiscal impact on the American taxpayer . . . may defeat us."

JSSAP seemed ready to meet that challenge. According to a spokesperson, the philosophy was a five-to-ten-year plan for combined U.S. military armament design and procurement. This original plan called for a memo of agreement, with all five services signing, for 500,000 XM9 handguns, 20 percent (100,000) of which would be supplier-suppressed, although all weapons must be suppressor compatible.

In addition to the pistol, JSSAP specified a requirement for both an individual and a sniper rifle, a submachine gun, a combat shotgun, a squad automatic weapon, and a heavy machine gun. Because there was controversy among the services as to just what the design specifications and the field capability of each weapon category should be, the plan has been slow to progress on schedule.

For example, the original plans for the individual rifle included a sniper model with suppressor. Did this eliminate the need for a separate sniper rifle? According to several insiders, the best bet for a sniper rifle was a slightly modified version of the XM21 of Vietnam-era fame. But could that also be an individual rifle? There was no general agreement on any model and even less on the standard service rifle, until the Marine Corps adopted the M16A2, which is a different story.

The 9mm submachine gun was supposed to utilize a closed-bolt mode for semiautomatic fire and an open-bolt capability for full auto firing. It was also to be suppressor-ready for easy conversion. One designer told me, "They want something along the lines of Ingram's M10, but a bit more sophisticated, whatever that might mean."

A combination of controversy, internal squabbling, ineptitude, indecision, and underfunding hurt the JSSAP concept. As 1983 came into being, the apparently stillborn programs listed the following status: XM9 pistol (in limbo); M249 FN minimi squad assault weapon (adopted); improved M16 (evolved into the USMC-adopted M16A2); advanced infantry rifle (contract awarded to HK for "an advanced concept," probably their caseless G11); close assault weapon, a euphemism for a combat shotgun (in limbo); general heavy purpose machine gun (probably to be JSSAP's own "Dover Devil" as a contract for the weapon was awarded to AAI, Inc., the folks who gave us the M73 and the M85); a suppressed submachine gun procurement program was in the advanced development stage; a sniper rifle program was in a sort of nonstart limbo; the long-range sniper rifle in .50 caliber and several other miscellaneous ordnance and related accessory projects were also in limbo.

Perhaps JSSAP managed to put a crack in the Ordnance Ring. But that remains a less significant achievement when one considers that the American arms industries are now facing the same problems being battled by domestic manufacturers in the consumer field, e.g., automobiles, steel, electronics, clothing, even mushrooms and potatoes. They are faced with aggressive, tough, imaginative foreign competition for the U.S. defense dollar.

For example, in the JSSAP XM9 trials, only two of the four pistols entered were American designs—those by S & W and Maremont. The kicker is that both of those companies are actually subsidiaries of larger parent companies in Europe. Of course, Beretta, FN, and HK are all producing weapons and parts in American facilities. But the cogent factor is that the innovation, design, and some financing is coming from Europe. How long will it be until more financing starts coming from the Middle East?

Perhaps it would be well to take a brief look at each of the JSSAP projects most appropriate to the general topic of this book.

The Pistol

This is the most ambitious project and the one that has generated the most publicity, both favorable and otherwise. It is also the area in which suppressors have gained the most attention.

On 6 May 1981, the U.S. Air Force Armament Laboratory at Eglin Air Force Base in Florida sent out a contract request asking companies and individual designers with experience and/or expertise in small arms suppressors to contact that military facility for possible business. Initial review would begin there for the 9mm pistols, all of which would have suppressor capability, according to Maj. John Toner of the Eglin lab.

The key use of these weapons would be as issue to combat air crews and pilots to increase their chances of survival if they were downed behind enemy lines. Maj. Toner said the suppressed pistols would also be Joint Service ordered for units "engaged in counterinsurgency and special operations." These types of operations might include "snatch missions involving prisoners or enemy VIPs or to seize enemy documents or intelligence materials behind the lines." Actually less than 2 percent of the planned 100,000 silenced 9mm pistols would be used for clandestine, intelligence, and special mission purposes. The vast majority would be issued for Air Force escape and evasion use.

One of the initial designs came from Eglin's own labs. A former research person at the U.S.

WILLIAM SEYMOUR

European law enforcement officials examine the HK MPSD, silenced submachine gun that has also scored well in JSSAP trials.

Army's Armament Research and Development Command (ARRADCOM) told me about the Eglin experimental suppressor that was rectangular, rather than round. It had the same width as the pistol and was frame-mounted no higher than the sights. It used the early Navy model 0 suppressor as the basic design for the system. This suppressor-equipped weapon was capable of either semiautomatic or single-shot operation.

The rectangular Eglin suppressor was a wipeless, off-center unit mounted low on the weapon's frame, using the Mark 22-0 performance as the baseline for noise level comparison. Although the suppressor was supposedly just a design concept, Smith & Wesson was awarded a contract to develop the unit.

An engineer who saw the concept called the idea "somewhat silly." He said it was designed for those pistols having a moving-barrel locking system, for example, the S & W, Colt, and Beretta. According to both my source and my own research, there are at least two existing suppressors that will do the same job. The major problem is making a suppressor unit light enough, about nine ounces, so as not to impede the barrel movement. Each of the two existing units weigh nine ounces or less and give outstanding noise levels, below 124 dB. Both are wipeless and one is eccentric, allowing the use of the existing pistol sight without modification.

This apparent duplication did not sit quietly among some critics who were sorry to see JSSAP management veer from their announced course of economy and nonpartisan decisions.

One critic, a former defense contractor who knows the suppressor field well, commented, "I don't know if the Eglin contract is money to soothe S & W over the XM9 business loss. But I do think they are simply reinventing the wheel up there (at S & W). There are a couple, maybe three, other excellent suppressors already in at least prototype stage that meet every specification. Why do they need to send thousands of taxpayer dollars up to S & W to replicate what someone else has already done and done well?"

S & W refused to respond to that question or my rephrasing of it. A spokesman at Eglin said that he had no comment beyond, "We are strictly a design and testing facility for the Air Force."

What about the initial pistol testing? How well did it support the JSSAP concept? The first round of testing was conducted at Eglin AFB late in 1981, and the Beretta 92SB was the favored weapon. The tests were conducted at Eglin since the Air Force was considered the prime service for the XM9 pistol. But following some behind-the-smoke-filled-congressional-anterooms lobbying by U.S. arms interests, JSSAP declared the Eglin tests biased. Several sources claimed that S & W had applied enough domestic political clout to stall any further move toward adoption of the Beretta. As an ironic feature, shortly after all this happened, S & W's parent firm sold control of the arms maker to a foreign corporation.

A second round of testing was held in May of 1982, conducted by ARRADCOM personnel using the HK P7, the S & W 459A, the Beretta, and the SIG/Maremont entry. Although the results are classified, most sources report that the HK model "won" this competition handily. However, the Department of Defense totally canceled the contract award. At least two of the four firms, Beretta and S & W, sued the U.S. government over the issue and the tests.

Although another round of tests were scheduled for sometime in 1983, observers were skeptical. As *Defense and Foreign Affairs Magazine* pointed out in its December 1982 issue, "Another round of tests is reportedly forthcoming, this time with a much larger number of candidates likely. But if the program couldn't manage a four-way competition, what are its chances of a ten-way one?"

Three of the four candidates for the XM9 JSSAP pistol trials were displayed at the United States Army Association (AUSA) show held in October 1981 in Washington. HK displayed their P7; S & W showed their 459A; Saco Defense Division (of Maremont) brought their SIG/Maremont P226. In addition, FN displayed their GP pistol, although it is not officially in the competition. All of these weapons are 9mm models with large magazine capacities and can be suppressor equipped, with modification.

In addition, although Colt had a stainless steel model in the running and displayed a 9mm sample of their SSP for the AUSA show, late that fall they said they would not enter the competition because of their deep involvement in the M16 improvement program, another part of the JSSAP operation. Colt did submit a proposal, however, to convert existing M1911A1 pistols to 9mm, a move which got great congressional and media support.

Even before the 1982 tests, the JSSAP XM9 project was in trouble, due to legal threats, interservice rivalry, conflicting product claims, political pressure, lobbying . . . all the nastiness that the JSSAP concept was supposed to eliminate. While you can theoretically take the old boys away from the action, you can't take the action away from the old boys.

When the XM9 proposal got to the check writers in the House of Representatives, money had become the key public issue. In October of 1981, the House Armed Services investigations subcommittee withdrew funding for the XM9 project after an Air Force gunsmith converted a 1911A1 to 9mm in ten minutes at a cost of less than one hundred dollars. The congressmen were being asked to spend $200 million to produce 500,000 XM9 weapons; that's $400 per pistol. The loyal opposition said the conversions could be had for less than a quarter of that.

By late 1982, the XM9 concept was in financial limbo, its procurement funding cut by Congress. At the time, the public debate seemed to center around the old hassle of .45 vs. 9mm or "what was good enough for the boys in 1911 is still good for the boys today."

Some of the research work went ahead, however, on a less ambitious basis. For example, at Eglin AFB, Jack Robbins, a respected small arms project supervisor, said, "We're doing generic research, testing toward a suppressor that will knock down the dB. We're looking for better designs to improve efficiency and for better ways to attach the device to the weapon."

According to Robbins, the Eglin facility had

contracts with Beretta and S & W to develop a generic suppressor that would work well with various pistol designs. "Basically, the S & W design is the Browning type, while the Beretta is the straight-back action. In December of 1982, we had just gotten started in finding a good, workable 9mm suppressor for these handguns. We're not looking to Flash Gordon or Star Wars . . . just good basic science and design," he added.

Robbins also said that the majority of the suppressed weapons would be for escape and evasion missions and would follow the criteria listed in the procurement requests.

Although the exact JSSAP specifications for a suppressor were not known beyond suppliers and others privy to classified information, the following requirements for an ideal military suppressor have been suggested by several inside sources close to the testing program:

- An acoustical performance showing a drop of at least 30 dB, preferably over 40.
- A life expectancy compatible with the host weapon, that is, tens of thousands of rounds.
- Total elimination of muzzle flash, even at full auto fire.
- Ease of assembly by field personnel.
- Little or no user maintenance.
- Complete interchangeability between issue weapons without special adjustment.
- Positive alignment undisturbed by combat conditions or rough handling.
- As compact and light as possible given the maximum sound reduction capability.
- No reduction in basic weapon accuracy and protectile velocity unless using subsonic ammunition.
- Simplicity and economy of manufacture using nonstrategic materials.

With these needs in mind, reportedly, two top suppressor designers, Reed Knight and Don Walsh, were asked by JSSAP officials to combine their talents and experiences to come up with a hybrid model suppressor, which they did in 1982. According to one insider at Dover, "We wanted something that combined the Knight wipe on a Walsh suppressor. This would lower noise and lengthen the unit's life. It would be about seven and one half ounces and purely for pistol application.

"At present, the Walsh unit we're testing on the handgun has a level of 126 dB, lots better than the 136 dB drop of the next unit. We think the Knight/Walsh hybrid will give us another 5 to 10 dB of sound reduction. . . . That's damned impressive."

The Knight/Walsh hybrid was superior in noise level reduction to Knight's own improved Hush Puppy when used on the HK P7, P9S, and the slide-locked Beretta 92. The new unit weighs less than the Hush Puppy and is slow to burn out—reach an unacceptable noise level—both of which are in its favor.

According to one JSSAP source, even the hybrid has been overshadowed by Don Walsh's second generation 1983 Larand suppressor, which is an eccentric model, wipeless, and light enough for operation on the moving-barrel weapons. The man related, "If nothing else, this Larand model will mean the end of the suppressor wipe. They are simply no longer needed."

Does this mean suppressor wipes are no longer needed on the windshield of sound vulnerability? Much to the dismay of purists and wipe makers/sellers, these devices will become as obsolete as disco music. The final nail in their operational coffin is the increasing popularity of high performance specialty ammunition. Both ultra-velocity, armor-piercing rounds and the prefragmented "safety slugs," such as the incredibly lethal "Glaser" developed by Col. Jack Canon, are incompatible with wipes. Most experts agree that within a few years no modern suppressor will utilize a wipe, unless it is a small, disposable-type silencing unit.

In addition, HK had been subcontracted for concept and feasibility research to add suppressors to some of its other weapons, including an assault weapon, a shotgun, and their belt machine gun. There was also some interest in suppressors at GE, Westinghouse, and Maremont. This was about the extent of JSSAP suppressor work as 1983 rolled into view. The program was resting uneasily on an inactive dead center, slowed by a drain of dollars.

Perhaps the most effective way to get the JSSAP suppressor program rolling again would be to have the U.S. Surgeon General declare that firearm noise pollution is harmful to the health of American military personnel!

However, as some work, at least at the conceptual level, was done beyond the XM9 in each of the weapons genres prior to 1983, some review of the various other JSSAP small arms programs would be in order.

The Individual Rifle

As *Defense and Foreign Affairs Magazine* reported in its December 1982 issue, "Any lingering doubts about the long-term U.S. commitment to the M16 system are about to be removed by the adoption of the M16A2 product-improved weapon." The U.S. Marine Corps adopted the "new" weapon late in 1982. Basically, the M16A2 has been given a heavier barrel with a faster twist to accommodate the new L110 and SS109 ammunition. There is a round forearm, a stronger stock, and a luminous sight. Many of the changes remind ordnance observers of the original design done a number of years ago by Eugene Stoner. As with its older models, the new M16A2 is suppressor-compatible.

The Sniper Rifle

Vietnam taught the U.S. military the value of effective sniping programs, so a major part of the JSSAP program involved sniping weapons. For example, at the government's Rock Island Arsenal, technicians tried to update the XM21 sniper rifle, which is basically a precision variety of the M14. There was little success in trying to mass produce these precision weapons which traditionally had been nearly hand-built by the gunsmiths at the Army Marksmanship Training Unit (AMTU) at Ft. Benning.

When JSSAP sought to gain an interservice sniper rifle, they found at least two major conflicting philosophies. The Marines favored their bolt action M40, while the Army liked its semiautomatic XM21. To add to the battle, two newcomers were included in the JSSAP plans. One was the HK PSG1, an updated variation of the G3 in 7.62 NATO, that had been modified for sniping work and included a suppressor and a silent bolt closing system. The price tag on this system is $5,500 per rifle. The fourth possibility was the Galil sniper model, a version of the IMI assault rifle in 7.62 NATO. This weapon was personally designed by Col. Alex Eliraz, the renowned and retired Israeli sniper, and is compatible with a suppressor.

Plans also called for a heavy-caliber, long-range sniper rifle, an old idea that still has supporters.

"One of the weapon designs was a very long-range sniper rifle in .50 caliber for which they wanted a suppressor. And, they aren't talking a crew-served weapon," one JSSAP ordnance officer related. "Sniper rifles in that caliber were used in Korea as field improvisations, and I know they were used on a limited basis in Vietnam, so it's not really a new concept."

The only official field testing I know of involved the Winchester Model 70 in .458. Suppressed by the Human Engineering Lab at Aberdeen Proving Ground, the weapon failed its field tests in Vietnam. The developmental history of this monstrous weapon is told in detail and with accuracy by Peter Senich in his fine book *Limited War Sniping* (Paladin Press). In summary, the only feature of the weapon that passed muster was its quieting capability. The weight, accuracy, and other mechanical functions were a nightmare. Yet according to insiders at JSSAP, this weapon and its integral suppressor have been refined at Aberdeen's Human Engineering Lab and have performed very well to date. Reportedly, one was taken along for field testing on the Iranian hostage rescue raid.

The Combat Shotgun

For years, many people have been under the impression that the use of the shotgun was banned in warfare under the terms of the Geneva Convention. But then, so are punji stakes, poisons, dum-dum bullets, Malay gates, and tossing prisoners out of helicopters. The irony is that the shotgun is one of the most lethal individual combat weapons around. JSSAP recognizes this and wanted our boys, and possibly girls, to have the very best. For that reason, JSSAP assigned a Navy small arms engineering team at the Naval Weapons Support Center to manage its Multipurpose Individual Weapon program—the combat shotgun.

The basic requirement was a versatile, selective-fire military shotgun. Considering that the Manville-type tear gas gun, as mocked up for that sexy weapon in *The Dogs of War,* really is available for a variety of 30mm projectiles, can such a device be far behind for JSSAP? Both Winchester and HK reportedly have a military shotgun design. At this writing, however, only HK has a prototype. People who have seen it say it resembles their G-11 caseless cartridge bullpup design. There are rumors that

A Marine expert test fires the Dover Devil 50 caliber machine gun—a weapon with potential for suppressor adaptation.

JSSAP personnel are also doing some R & D work on the feasibility of a silenced shotgun, a project undertaken twice during the Vietnam years.

In terms of ammunition, a variety of loads, both conventional and otherwise, exist in 12 gauge, e.g., explosive, saboted slug, flechette, nerve gas, incendiary, etc.

The Submachine Gun

The fate of the U.S. military submachine gun is about as settled as Elizabeth Taylor's marital status. During JSSAP's operating days, the only practical interest came from the special warfare units which have a tactical need for silenced weapons. What remained in the U.S. arsenal included a number of World War II era M3A1 "Greaseguns," plus inventories of the Uzi, Ingram or MAC M10 and M11 models, the S & W M76, and the XM177E2 variant of the M16, and the AR18, which are not true submachine guns.

According to most observers, JSSAP held a strong prejudice against the typical open-bolt slam-fire action. The only major weapon which avoided that was the HK MP5 and its silenced version, the MP5SD. In addition to the military, a number of intelligence and security agencies use the MP5SD, as well as its newer version, the HK 54A1, which has an adjustable gas porting system to accommodate both normal ammunition and the subsonic 9mm ammo used in suppressed operations. A prototype HK 54A1 was purchased by the U.S. Navy in 1982 for $90,000. Reportedly, Heckler and Koch claimed an R & D expense in excess of

$400,000 for this new weapon.

This raises the question of cost. The HK weapon is an expensive one, while one of the major arguments for the submachine gun is that it is supposed to be a simple, highly automated, and very inexpensive weapon to manufacture, e.g. the Sten, M3A1 and M76.

Although submachine gun R & D was moribund in the U.S. at the time, several firms were doing work on these weapons, e.g., the Saco Defense Systems Division of Maremont Corporation. Guilford Engineering Associates performed an R & D project for the Navy involving a novel design utilizing a moving barrel which negates both climb and recoil. Another similar submachine gun was successfully demonstrated at Ft. Benning by Jim Leatherwood of Leatherwood Brothers. He's the genius who invented the Leatherwood Auto Ranging Telescopic sight. According to observers, the weapon had definite Ingram lines, understandable as Leatherwood was once chief engineer for Military Armament Corporation.

Three existing miniature submachine guns that should have attracted attention within JSSAP were the Ingram M10, the M11, and the Mini Uzi, excellent weapons for suppressors. The Mini Uzi is a real favorite of the intelligence, law enforcement, and special mission warriors because it combines all that's wonderful about the Uzi with the compact size associated with the Ingram weapons.

Rumors were strong in 1983 that Action Arms of Philadelphia, importer of the Uzi and Mini Uzi, was going to produce a submachine gun called the Universal Machine Carbine, which is suppressor-compatible. As this book was going to press, however, details were unavailable.

The Squad Automatic Weapon

The new XM249, aka the FN Minimi, is the new JSSAP squad automatic weapon, filling in for that wonderful old veteran, the venerable BAR, retired in the early 1960s. In JSSAP testing, the FN Minimi won over a weapon submitted by Ford/Rodman Labs, one by HK and a modified M16. The XM249 is in 5.56mm and uses the Belgian SS109 and L110 ammunition, including an improved daylight tracer round and a semiarmor-piercing projectile. The weapon accepts both M16 and FNC magazines as well as 200-round belts. Several individuals and firms were approached about providing a sound suppressor for the XM249.

Maremont came out with a semicompetitor squad weapon to please the 7.62 fans—their lightweight version of the M60. At 18.5 pounds, the M60LW is a fully controllable 7.62 NATO belt-fed gun. Maremont also instituted a feasibility study to add a suppressor system to this weapon.

Heavy Machine Guns

There is, of course, only one true .50 caliber machine gun as any good soldier knows—the great old Browning M2, still on active duty. However, JSSAP had a better plan, as did Maremont. The latter built a lightweight infantry version of the M2, while JSSAP has its own .50 caliber weapon called the "Dover Devil" because their Projects Office is at Picatinny Arsenal in Dover, New Jersey. The Dover Devil, which is capable of being modified for suppressor use, is an interesting design, weighing half of the M2's mass and using only half its parts. It also features a dual feed so that two types of ammunition can be fired selectively. It is a very versatile weapon, too. Changing a few parts converts the weapon from a .50 caliber machine gun to a 20mm cannon, suitable for engaging hard skin targets.

9. The Terrorizing Sound of Silencers

Clammy moisture formed on the upper lip of the security chief and his stomach felt queasy. Only a few more minutes, he thought, and the damn OPEC people will board their jetliner and become someone else's problem.

Primal fear lifted the hair at the nape of his neck just before the first round snapped within a foot of his head and thudded heavily into the Libyan oil minister. Another bullet cracked and hit the Algerian representative.

In a frenzy, the security chief whirled around searching for the sound of thunderous gunfire inside the airport complex. He had heard none, only two soft pops from a balcony.

"Silencers!" he roared to his detail. "They're using bloody damned silencers."

A few moments later, one of his men recovered the abandoned murder weapon, a Remington 700 with a long, thick tube on the end–a professionally made silencer.

Although the scenario is highly realistic, this scene hasn't happened, yet. Consider: terrorists wishing to blame the American CIA in the eyes of the Third World locate a well-known U.S. sniper rifle, put a highly efficient silencer on the weapon, shoot two OPEC officials, then leave the weapon behind for local police and media to find.

Don't smirk. Our CIA and their KGB do things like that to each other all the time. Now, though, a third trigger finger has been added to the game–international terrorism. Terrorists have discovered the science/art of quiet death: firearms silencers. In an attack first pinned on the Israeli Mossad, two PLO officials were killed in Nicosia, Cyprus late in 1979 when a masked gunman armed with a 9mm silencer-equipped pistol calmly shot and killed the two men less than one hundred yards from a police checkpoint. Neither officer on duty heard the gunfire. Barring some police calumny, that sort of field tested efficiency is what quality silencers are all about.

In their typical mist of black propaganda, the Israelis remained totally silent about the killings. By early 1981 though, police officials in Nicosia confirmed that the murders were the work of potential rivals to Yassir Arafat within the Arab world. The weapon used was a Beretta equipped with a "well-made, professional silencing device made in Eastern Europe," according to officials in Cyprus.

The Mideast murders continued. In July of 1980, former Syrian premier Salaheddin al Bitar was assassinated with a silencer-equipped handgun. Then, as a sidebar to the Iranian hostage story, in 1980 Paris police arrested five members of what they described as "a well-trained Islamic hit squad" whose target was Shahpour Bakhtiar, the late Shah's prime minister. Bakhtiar, an outspoken opponent of the Iranian revolutionary leader Ayatollah Khomeini, was living in France. A team of gunmen tried shooting their way into the Iranian's fortified apartment with silenced Beretta pistols. A policeman and a woman neighbor were killed in the wild exchange of shots, during which two of the five terrorists were wounded–obviously, a case of Khomeini grits.

In the Philippines, terrorist/bandits operating under the guise of Moro freedom fighters attacked a government bus using a silenced AK47 to take out two of the military guards riding escort. An

American passenger who saw the weapon in action was a Vietvet who could easily identify the Kalashnikov but who had never seen a silenced model before.

"These random terrorist attacks are not new," said the late Fred Stock, a former CIA field officer who was very familiar with the use of silenced weapons on special missions. "What's new is the increased silencer technology. It used to be that once in a while you'd hear about a homemade can on the end of a weapon. Today, these people have access to the latest and best equipment available."

There is documented silencer use among the New People's Army, a Mao-aligned group in Luzon, according to sources in the Philippine military. However, further south, more silencers turn up among the troops of the Moro National Liberation Front. This nationalistic group of freedom fighters uses M16s and AKs, fabricating silencers in their own workshops just as their fathers did in WWII for use against the Japanese. Also, the urban splinter group run by the Lovely brothers, which was broken up in 1981 by the Philippine Constabulatory, had targeted individual government officials for assassination using their stockpile of silenced weapons. Authorities not only seized operational plans, but also broke up an arms buy which would have added two silenced Skorpions to the brothers' arsenal.

Obviously, silencers are a favorite tool of terrorists who come by their silencers via the domestic black market, theft from the military, and the purchase or gift of silencer-equipped foreign military and intelligence service weapons. Unlike the U.S., several nations permit the open sale of silencers. Thus, the devices are legally bought in one nation, then illegally smuggled into other countries where they are sold to terrorists and criminals at a fat markup. A silencer bought in France for $30 may bring $750 in the U.S., and even more in Ireland.

When Pierluigi Concutelli, a chieftain in the Italian terrorist group Ordine Nuovo, was arrested in 1977, police also confiscated his silenced Ingram M10. It was the same weapon he'd used to murder an Italian magistrate and a police investigator who was closing in on Concutelli's Mafia ties.

Spanish terrorist groups like Ingrams, too. A sizeable consignment of M10s, sent under the auspices of official U.S. authority to DGS, a Spanish intelligence agency, somehow ended up in the hands of domestic terrorists.

There are easier ways to obtain silenced weapons. The Communist bloc is more than generous with their ordnance as long as they're certain it won't backfire. Both the Soviet Union and Communist China have already donated their top-of-the-line assassination weapons to terrorists.

Back in 1977, an American intelligence agency paid $25,000—no questions asked—for a sixteen-year-old Soviet military rifle. At the time, this was a weapon so mysterious and well guarded that no picture even existed in the West until then. KGB assassins, Soviet military snipers and terrorist gunmen used this super sniper rifle to kill American diplomats, intelligence agents, and military officers in the ongoing cold war under the surface of detente. The weapon is known as the Dragunov.

The captured Dragunov was turned over to the U.S. Army's Foreign Science and Technology Center (FSTC) for study and testing. According to an FSTC technical expert, the Soviet weapon is "a

JOHN A. MINNERY

This 20mm Lahti model 39 antitank cannon was fitted with a homemade oil drum silencer and used to blast open a vault in Quebec.

JEAN RENE SOUETRE

Terrorist trainee with Sterling using fabricated suppressor made in sophisticated armory setup. Unit is anodized aluminum with both baffles and packing.

CHRIS TRUBY

Author with a Sterling 9mm submachine gun using a field expedient suppressor designed for terrorists in the Middle East in the late 1970s.

hybrid semiautomatic sniper rifle which rates near the top of high-quality match weapons."

An Army Special Forces officer who handled the Dragunov says, "What makes this weapon so deadly is its extreme accuracy at ranges of up to a mile. Plus, it is semiautomatic, which means a sniper can kill several people within a group, just by quickly switching from target to target after each shot.

"The telescopic sight is also automatic. It ranges for the shooter, so that all he has to do is line up the sights and pull the trigger. The gun does the rest.

"Finally, a silencer can be attached to the rifle to make it virtually undetectable at ranges of 200 meters. This makes it one damn fearful weapon."

Even more chilling is a report by the British journalist Otto Oldfield from the Middle East that the PLO has these rifles. "PLO assassins are armed with this high-powered, super-accurate Soviet sniper rifle that can kill at ranges of nearly a mile," Oldfield writes. "This sniper rifle is so well designed that when it is fitted with a silencer, it can be fired without detection at fairly close ranges. Beyond 300 meters it is virtually undetectable."

Oldfield also writes that a silenced Dragunov was used to kill the American consul in Cyprus during the 1974 fighting there. "The word I got was that the Soviets gave out some of these rifles even before a specimen fell into western hands. It is a damned scary business, knowing PLO hitmen have these highly technical sniper weapons," Oldfield adds.

"I can't prove it, but I am certain that the Dragunov has been used to eliminate western agents, including Americans. The Agency (CIA) won't confirm it, but I personally know of one American citizen who was killed in the line of duty over there (the Middle East) a year or so ago (1977) by a PLO sniper firing a silenced rifle. The slug the locals took out of his body is that exact same type and caliber used by the Soviets in their Dragunov. Your chap was working as a contract employee for the Agency," Oldfield reveals.

Not to be outdone by their bitter Soviet rivals to the west, the Chinese Communist ordnance

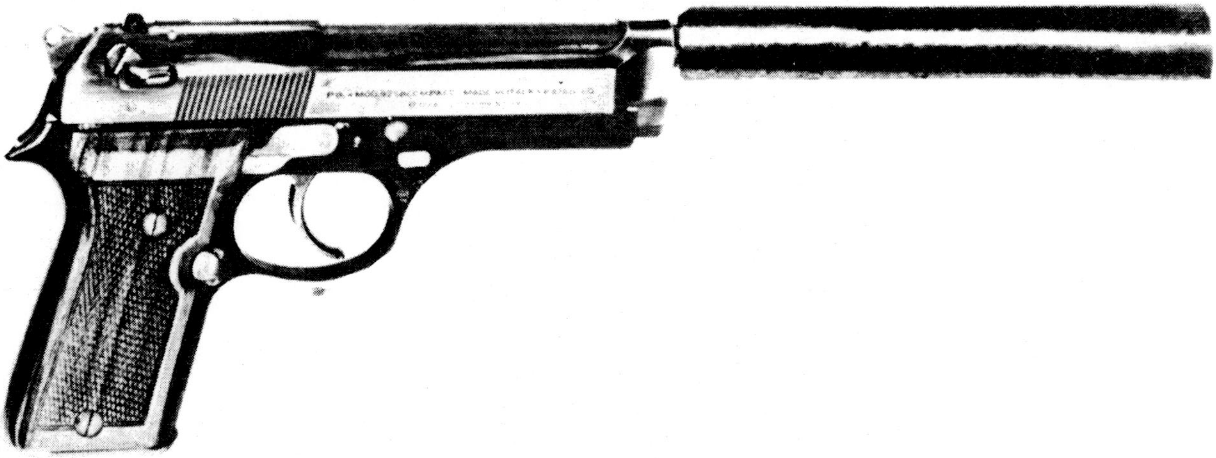

EMILIO SANTANA

This Beretta 92 in 9mm was recovered by the Italian police from the terrorist group that kidnapped U.S. General Dozier. It has reportedly been shipped to the U.S. for test and evaluation as the design is "quite decent for field work," according to a transmittal report.

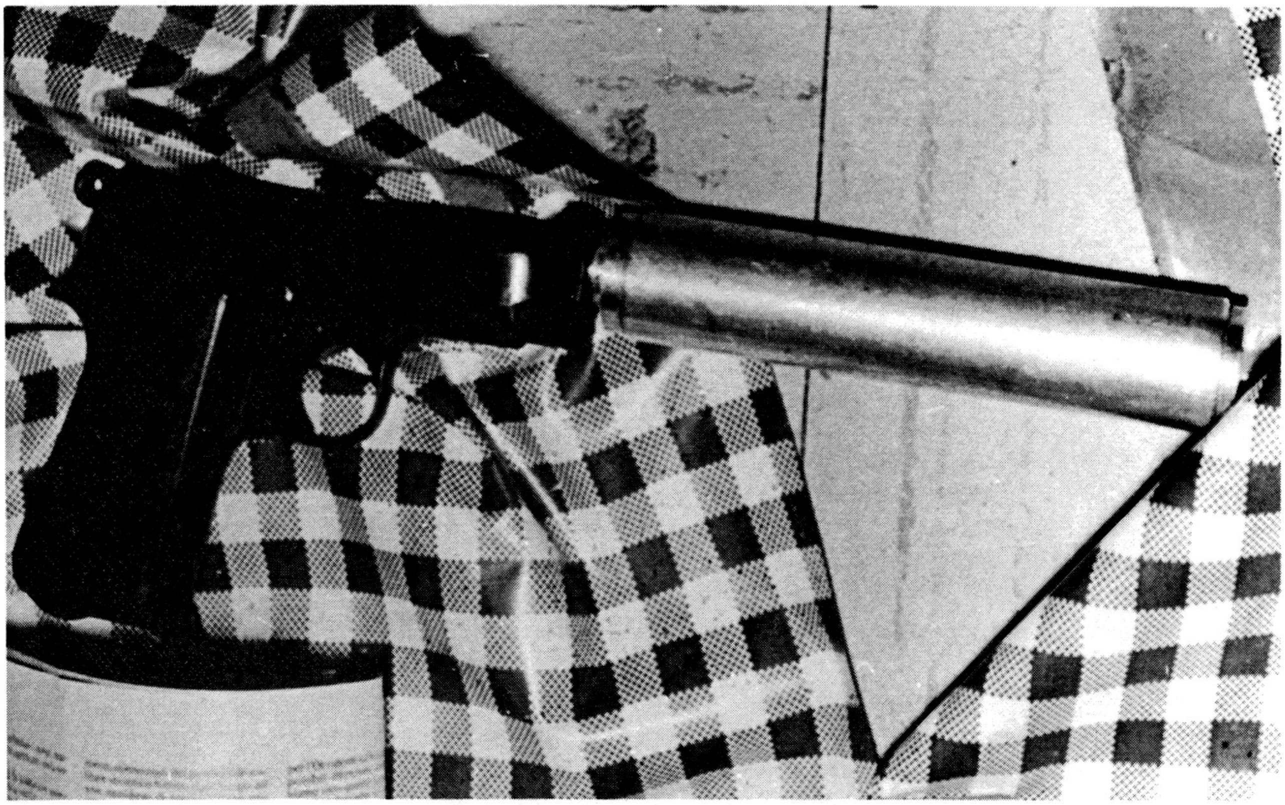

GIANFRANCO CORRIAS, QUESTORE DI PADOVA

The 9mm Beretta used as principle weapon in the kidnapping of General Dozier in Italy, 1982.

people have an efficient assassination weapon, too, the Model 67 pistol. According to a recently declassified after-action report from the Defense Intelligence Agency, the first of the ChiCom Model 67 pistols showed up in Vietnam in 1968. In addition to reporting this pistol at an NVA encampment, the "unit asset" as the Americans called their agent, also reported that she saw a silencer-equipped Skorpion and some sort of Polish submachine gun with silencer.

Issued only to top agents in 1968, the Model 67 was carefully secured in field use for seven years. American authorities finally secured a specimen during our final days in Vietnam in 1975. Since that time, specimens have found their way into other hands—unfriendly hands with busy trigger fingers.

In 1980, two terrorists tried to cancel the ticket of one candidate in a local election in India using the silenced ChiCom weapon. Alert police wounded one of the assassins, but the man with the weapon escaped. Then, early in 1981, Israeli commandoes recovered another of these silenced ChiCom pistols after a flurried firefight during one of those series of small raids into Lebanon on Operation Strikeback. The Model 67 had been in the possession of a PLO professional who had no further need for it since he was dead.

"This is an assassin's gun—close-in and silenced, a weapon for a professional killer," according to the late Fred Stock. "There is no conventional military use for this weapon. It's for silent murder."

According to one official field test report thought to be CIA in origin, the weapon has about "one-fourth as loud a report as an ordinary .22 cartridge in a rifle and less than a cap pistol." The weapon fires a special round, the 7.65 x 17mm, a rimless cartridge unique to this pistol.

The FSTC evaluated the ChiCom weapon by noting, "The Type 67 silenced pistol appears to be well designed and the materials and workmanship consistent with that of U.S. military weaponry."

A U.S. Army ordnance officer added, "This pistol is solid and well made, contrary to our usual propaganda about their ordnance and equipment. It's produced from machined-steel components and plastics. The silencer is a modification of the original Hiram Maxim design mounted concentric to the vented barrel.

"The guts of the silencer are a steel tube partially filled with a roll of copper mesh screening fitted over the barrel. The rest of the silencer housing is filled with steel baffles and rubber washers to form a series of small expansion chambers. It's a damn fine design.

"Firing from a nine-round magazine, this pistol is selective fire. You can switch from regular semiautomatic to the really silent mode of a locked bolt, in which single-shot, silenced gunfire is possible. This manual operation shuts down all the bolt noise and the loud escape of residual gases caused by the cartridge explosion."

Tested against one of the finest silenced combat weapons in the free world, the British Welrod of WWII fame, the ChiCom Type 67 did well. "It was much more accurate than the British assassination pistol and very nearly as quiet, based on the extensive tests run by the Army," noted the Army ordnance officer, who added that the silenced Welrod test-fired at an average sound level of 117.4 dB, while the Chinese pistol registered 122.5 dB.

The two giants, China and the USSR, don't own the total market on quality silenced weapons, though. The third partner in this deadly Communist trio is one of the more esoteric weapons favored by our terrorist opponents—the Czech M61 submachine pistol, better known as the Skorpion. This 7.65mm weapon can be used as a pistol or as a mini-submachine gun in either semi- or full auto mode. It comes with a very effective silencer as an accessory, and many experts consider it the ideal weapon for the terrorist who wants something small with both full auto and sound suppression capability.

According to the late Fred Stock, some of these quieted killers from Eastern Europe showed up in both South and Central America. He told me, "A few specialists have been sighted with silenced Skorpions. I would guess they are coming in from Cuba."

In addition, according to William Chapman of the *Washington Post* foreign service desk, South Korean officials have interdicted several "terrorists and agents" in and near their coastal waters. The agents, presumed to be North Koreans according to Chapman, were armed with the silenced M61 weapons.

No respecter of age, rank, or retirement, the ubiquitous silenced Skorpion was used to murder Aldo Moro, the ex-premier of Italy in May of 1978. His Red Brigade assassins used a silenced Skorpion to kill Moro after holding him captive for fifty-four days. One of his killers was a woman terrorist.

In a dawn raid on an apartment near the Vatican in Rome, police rousted a Red Brigades couple, named them as Moro's assassins, then displayed a veritable arsenal of ammo, explosives, fake IDs, as well as weapons, including several silenced Skorpions and five semiautomatic pistols equipped with silencers.

Police said that after killing Moro this couple had raided a central Rome office of the Christian Democratic Party, shooting up the place and killing two policemen in the getaway. Witnesses said that the silenced Skorpions were used in the attack.

In one terrorist training camp observed by an agent working deep cover for a NATO country in 1980, selected Arab gunmen and women were learning how to use the silenced Skorpion for quick, quiet, and messy assassination hits. By the way, the camp was personally hosted by the charming Muammar Qaddafi in Libya, according to the NATO report I saw.

According to Mike Browning, a former State Department employee, a lethal gimmick reported in one of my earlier books was utilized by Edwin Wilson, an ex-CIA op who later became a Boswell to Qaddafi, the Libyan dictator. Browning says that Wilson saw my reportage of the silenced gun-in-a-camera assassination device and sought a contractor to build the device. He ordered 300 of the weapons, specified as two-shot .22 caliber firing devices with suppressors, the entire apparatus being built into the bodies of 35mm cameras equipped with telephoto lenses.

"Wilson got the devices, and they worked," Browning told me. "They were amazingly accurate for their size and very quiet. Each camera gun was a very lethal assassination device."

There is no way of knowing where any of the devices might be now, except that according to a published report by columnist Jack Anderson late in 1982, all of them were delivered to Qaddafi. Anderson writes, "There could be upwards of 300 of these assassination cameras currently in the hands of terrorists around the world."

The complexity of our modern society fits right in with terrorist battle plans. As the terrorism scholar Walter Laqueur puts it, "One hundred fifty years ago if someone wanted to put out all the lights in a village, he had to go from house to house to do it. Now, he blows up one generator and all the lights in the city go out."

10. Omerta at the End of a Gun

He looked as if he had as much business being a pro shooter as Bill Casey had running the CIA. A middle-aged, soft-middled man of 49, Clay Alexander isn't even his real name. Only his former employers and a few officials in the Federal Witness Relocation program know that. Indeed, Clay Alexander is a name we decided upon for my interview with him. As he said with a smile, "Some of my former peers will recognize the reference."

Alexander has owned up to eight shootings in dealing with the feds to stay out of prison. Three involved .22 caliber weapons equipped with suppressors. He is only one of four of the infamous ".22 caliber killers" in custody to date. They were a loosely based squad of about fifteen professional killers, including two women, who used .22 caliber handguns almost exclusively.

"Most of the pistols were equipped with silencers that we either bought on the black market or had a machinist make up for us," he related. "At one time I had a Beretta and a Taurus modified for silencers, and I used both of these weapons on the job. They were very efficient and quiet. Most of my jobs were in public or semipublic places, although not when there was a crowd. Still, gunshots make a helluva noise, so we used silencers.

"There's no such thing as a real silencer. But two of mine cut that blast in half at least. And even a .22 makes a sharp crack."

Like a true professional Alexander is sophisticated about the tools of his trade. He knows the principles of suppressor design and can compare the differences between baffle and mesh systems, ported versus unported barrels. He also told me he had read my first book on silencers. I'm not sure that I should feel flattered.

Only one of his silenced weapons was recovered with him, the Taurus. It tested as efficiently as the suppressed weapons used by government agencies and the military. All Alexander would say was that "a good machinist with some previous experience in silencer designing made it for me. He did a damn fine job, a real craftsman. Too bad he doesn't work for the Army, too."

The Taurus suppressor is about eight inches long, a bit more than an inch in diameter, and uses a standard baffle arrangement in combination with copper mesh screening. The weapon is inside-threaded for the unit and the suppressor's finish is quite professional.

According to cultural myth, silencers are as common to organized crime as celebrity gossip is to tabloid newspapers. Actually, that myth is now reality; silencers are routinely used by criminals, although *Omerta,* the underworld's code of silence, makes information tough to come by.

But one thing is certain: cracks will appear in the wall of silence around Mob murder almost as readily as potholes will appear in American highways. For example, one empirical witness and user who later documented the Mob's growing use of silenced pistols was Aladena "Jimmy the Weasel" Fratianno, who described many long-hidden secrets about organized crime hits to a federal court jury in Los Angeles late in 1980. Fratianno described several hits using silenced .22 pistols.

A physical break happened in September of 1980, when ATF agents seized silenced weapons in the home workshop of a former Chicago police officer. Early news reports in the *Chicago Sun-Times* said officials were concentrating on the possible links between the .22 pistols being silenced there and those sold to contacts for Mob hitmen.

A spokesman for the Chicago ATF told me, "We are investigating possible connections between many unsolved, crime-related homicides using silenced .22 pistols and this case. Two suspended city policemen are under investigation for possible manufacture and sale of these illegal weapons to potential perpetrators of such crimes as assassination."

The *Sun-Times* named former police officers Richard Madeja and Joseph P. Ahrens as the two men under federal investigation in the matter. The silenced weapons were found in Madeja's home workshop, while Ahrens was named as the alleged salesman.

The ATF spokesman said that bullets passing through a silenced weapon pick up markings distinguishable from normal land and groove deformations. The types of silencers produced and marketed in Chicago and in the seized weapons made the same distinctive marking as the markings found on the bullets which killed twenty organized-crime victims of the .22 caliber killers. He added that one of those victims was Mob boss Sam Giancana.

People who say hitmen are found only in the movies never met Bernard Barton Hunwick of Ft. Lauderdale. Described by police as "one of the biggest hitmen in the nation," Hunwick was arrested in 1982. A search of his lavish canal-front home turned up a true arsenal of weapons, the tools of his trade. Included were handguns, rifles, shotguns, two pounds of C4 explosive, and two of the famed hit kits—Ruger pistols with integral silencers. "Damn fine weapons, real quiet and professional," said Detective Sgt. Dale Adams of the Ft. Lauderdale police.

The hit business is an equal opportunity employer, too. In 1980, NYPD officers arrested a twenty-one-year-old woman who was by all definitions a *mechanic,* a professional killer for hire. Blanche Wright was accused of murdering four men and one woman in her brief career. After her arrest, she admitted to two additional hits. Wright told police that her favorite weapon was a silenced .22 pistol, because "it made so little noise, I was able to concentrate without flinching."

Her partner, Robert Young, also a professional killer, used a silenced pistol regularly in his work. Isn't it interesting how well professional gunmen flourish in New York City, the self-professed home of the nation's toughest gun law? The Mob isn't very impressed with New York City's fabled Sullivan Law, the gun-hater's legislative delight. Early in 1981, agents of the NYPD and the BATF conducted a joint raid which netted thirty-five professional bad guys plus a large assortment of submachine guns, assault rifles, explosives, silencers, and silenced weapons. One cop who was in on the raid told me, "It was like capturing part of an enemy army and their arsenal. These mobsters had M16s, greaseguns, Thompsons, a couple FNs, and a whole bunch of loose silencers to fit the various weapons. They also had some pistols and a couple

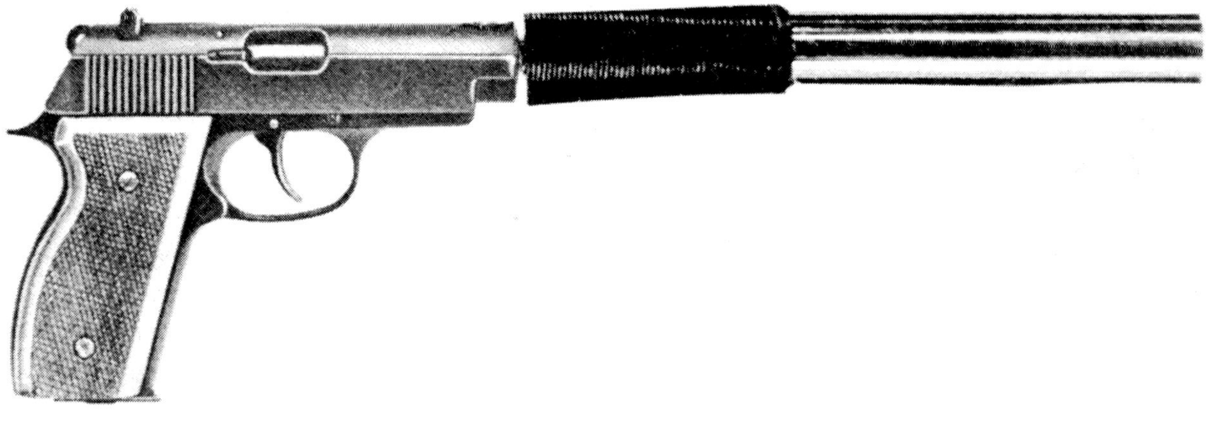

GORDON ALEXANDER

This Sterling 402 was recovered with its heavy, but useful suppressor late in 1982 from a felon with two prior arrests for assault with a deadly weapon. In testing, the suppressor caused a drop of .22 dB.

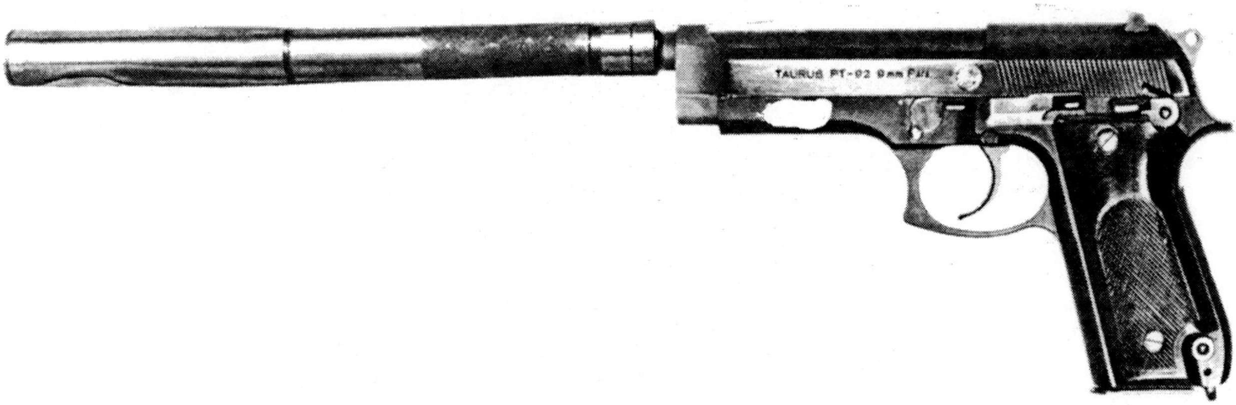

J. DAVID TRUBY

One of the Cocaine Cowboys carried this Taurus 9mm model PT92 with inside-threaded suppressor, until federal Strike Force people in Miami appropriated it from him along with some of his product. The weapon may figure in at least two shootings in southern Florida. Tested, the suppressor produced a drop of 26 dB. The serial numbers have been whited out to protect the cover of a source.

submachine guns with the silencers built right on them. All the silencers looked like they came from a professional factory. And this wasn't a major raid by any means."

In Newark in 1982, two brothers who were reportedly once members of a Hell's Angels gang were arrested with four cases containing 200 professionally made silencers. According to an ATF agent assigned to the case, the brothers, James and John Stevens, had intended to sell the silencers to two men they thought were underworld executioners. The "hitmen" turned out to be federal undercover cops.

According to George Schneider, the Essex County, New Jersey prosecutor, the silencers were manufactured in Denver then shipped to North Carolina. The two brothers transported them to New Jersey for sale into the metropolitan New York underworld marketplace. An ATF source said the illegal silencers dropped the blast 20 dB on several .22 pistols tested with them. Each unit is six inches long, one inch in diameter, and constructed of PVC pipe and hardened steel packed with steel wool and washers. Their street cost was estimated at $600 per unit. The 200 silencers were shipped to the U.S. Navy for testing and use, according to the ATF source.

In a 1982 bust, the LAPD broke its largest cocaine seizure ever with the note that a silenced Ruger handgun had been recovered as well. According to LAPD Cpt. Robert Blanchard, the weapon was top quality, both externally and internally. He added, "It was the work of a professional, that's for sure."

Dope and guns are the one-two punch of organized crime. Both are highly profitable indus-

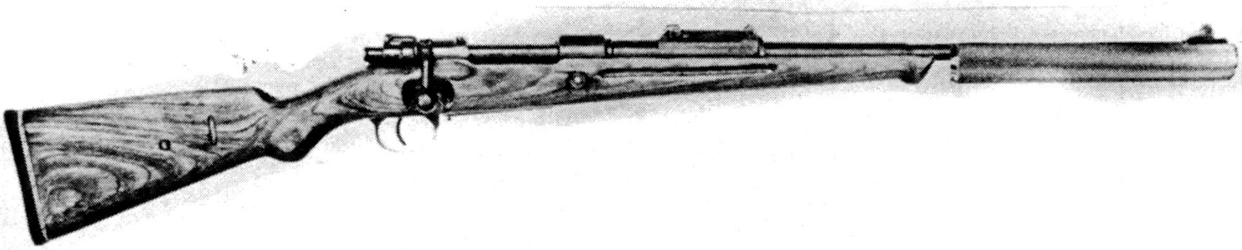

WILLIAM SEYMOUR

This wonderful antique, an old Mauser 98 in 7.92mm, was used by a Mob hitman to scare the credibility out of a witness in a drug case in one of our western states. Although the feds haven't tested the weapon, the man who made the suppressor says it will drop the noise by 25 dB.

J. DAVID TRUBY

A New Jersey heroin dealer lost this prize when federal folks shut down his supply operations. Tests showed a drop of 28 dB with this unit attached.

tries with the majority of activity in our southern states and below the border. That's where you might find Robert Vesco, the infamous friend of our infamous ex-president. Vesco's private army of Cuban exiles seems to have easy access to silenced Ingram weapons. Indeed, it's common to read that shipments of silenced Ingrams being sold to Latin American police and military forces have suddenly disappeared, probably under the time-worn Latino policy of to the highest bribe go the goodies.

However, not all the silencer movement is in ripped-off or other small lot quantities. During the middle 1970s, for example, the Cuban exile Alberto Sicilia-Falcon, allegedly the heroin czar of Mexico, was negotiating through a respectable intermediary with an American firm for rights to fabricate fully automatic weapons with silencers for some sort of clandestine action in Latin America. Knowing Sicilia-Falcon's reputation, it is not difficult to imagine the nature of that action.

The silenced weapon is not only a useful business accessory, it is also a status symbol for those who run the heavy drug traffic between the U.S. and Latin America. The media regularly carry stories of the dope-trade gang wars in which unemployed terrorists, thugs, freelance gunmen, and some old pros rampage through South Florida's urban civilization with silenced submachine guns. One report told of an "armored war wagon" found by police after several "missions." Abandoned during hot pursuit, it contained a 9mm Browning pistol with silencer, a silenced M3A1 submachine gun, an M1 carbine, two cheap imported revolvers, a .380 Beretta with silencer, and several ammo cases crammed with cartridges for the weapons.

Asked about the heavy traffic, one Justice Department official told me, "Our people and the ATF come up with only a tiny fraction of what is going down in illegal trade for automatic weapons and silencers. When you look at those numbers involved in seizures, remember that's only a drop in the bucket. That should scare the hell out of you."

He's right.

In April of 1981, federal agents in Miami recovered more than five hundred silencers in Operation Sky Drop which was an antidrug/anti-Mob project. A Miami police office said the silencers were "smooth, finished, professional designs . . . intended for either pistols or small submachine guns."

I asked him if that meant the Ingram M10. He shrugged, patted his .38 caliber issue pistol and replied sadly, "I hope to hell not. But, if so, I pray the bad guys sell them out of this country."

Things got so far out of hand with Miami's Cocaine Cowboys and their Wild West shootouts that late in 1979, the FBI added Mario Tauro Coto, a Cuban narcotics dealer and freelance shooter, to their famed illustrated poster list of ten biggie bad guys. This same poster later graced the *FBI Bulletin.* A man with a list of aliases as long as his yellow sheet, Mario Tauro Coto's favorite weapon is a silenced .22 caliber pistol.

These southern Florida drug wars of the 1980s must remind retired cops from Chicago, Detroit, and New York of Prohibition era blastouts. It's wild down there; it really is. In the fall of 1979, for example, nearly a quarter of the 250 murders in Dade County were related to the drug industry. Those deadly numbers grow steadily each year. These are basically territorial wars between the Colombian and Cuban drug wholesalers and their dealers. Dade County law enforcement officials dubbed the groups "The Cocaine Cowboys" and reported the use of heavily armored and armed dope-running vans which they called "war wagons." The favorite armament for both groups is the silenced M10 submachine gun, easily obtained with drug money on the street market.

In the spring of 1981, ATF agents recovered 1,460 silencers in Texas, Georgia, and Florida in unrelated incidents. The scenarios involved drugs, weapons selling rings, professional gunmen, and Mob workshop silencer factories. In Houston, for example, on 6 May 1981, undercover agents penetrated a silencer selling gang; three men waiting at the Houston airport expected to swap 800 silencers for $100,000 in cash. Instead of getting cash, the men were handcuffed and arrested.

In a guns-for-drugs deal in Georgia during "Operation Flying Circus," ATF agents recovered 620 silencers, plus twenty-nine machine guns, an airplane and 2,700 pounds of explosives.

In New York that same spring, ATF agents worked with the NYPD to break up a weapons smuggling operation with roots in Kentucky. "Operation Bluegrass" netted 129 illegal firearms and twenty silencers.

In another haul, ATF agents in Virginia and Maryland seized more than seven hundred weapons and sixty thousand rounds of ammunition. Included in the cache were submachine guns, silencers, sawed-off shotguns, hand grenades, mor-

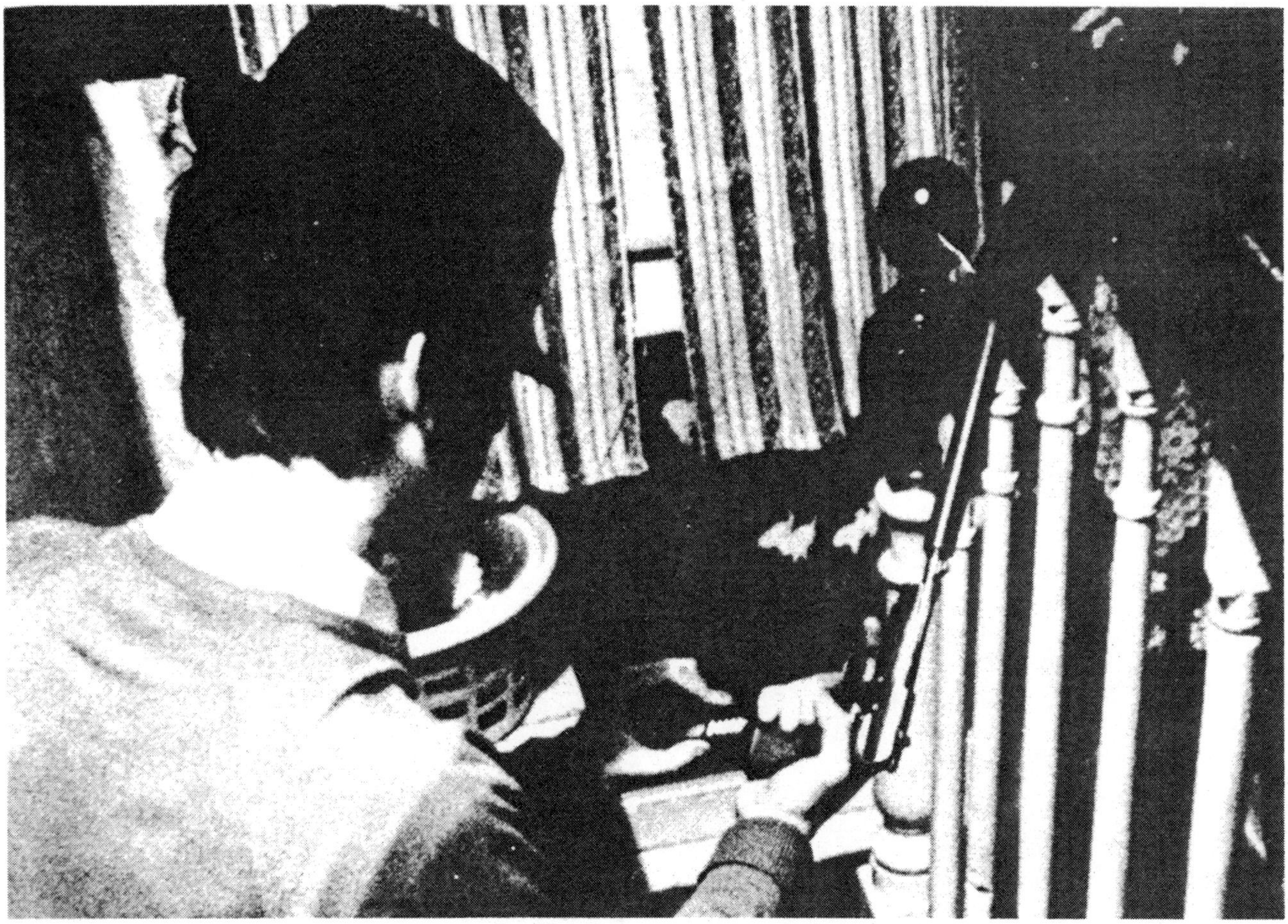

MICHAEL V. MERTZ

Actual photo of professional shooter with well-silenced .22 pistol, the real thing. Don't even ask. . . .

tars, and rocket launchers. In describing the raid, agents were very careful to tiptoe around the issue of who was to get these illegal weapons before the bust. Les Stanford, an ATF spokesman, said, "In our judgment, they (the weapons) were available to anybody with the money."

He added that the pattern of sales did not indicate that any specific groups were buying large numbers of the illegal weapons. But as former UPI newsman Bob Russel points out, "Just how large a number of machine guns or rocket launchers does a small terrorist cell have to buy to be considered significant?"

In the U.S., two groups come readily to mind when terrorist practice is mentioned: hardcore, outlaw bikers and the Ku Klux Klan. Bikers have always loved silencers and evidence shows they use them both operationally and for show. In their book *A Wayward Angel,* former Hell's Angel George Wethern and Vincent Colnett document various police seizures of Hell's Angels' arsenals during the 1960s and 1970s. Inventory lists include the usual submachine guns, assault rifles, pistols, light machine guns, grenades, dynamite, at least two .50 caliber machine guns, and a number of silenced weapons and silencers. Former Angel Sonny Barger not only had several silenced pistols, but according to Wethern and Colnett, also possessed one of the old, original Maxim silencers. They noted in their book:

> Not only was the president (Barger) interested in powerful automatic weapons for police trades, but he also sought out untraceable pistols of small caliber, mainly compact and easily silenced .22 and .32 automatics, the kinds of guns commonly used in professional killings.

In the spring of 1983, several former Hell's Angels testified under the Federal Witness Protection Program before the Senate Judiciary Committee investigating connections between bikers and organized crime. One former gang member known as Butch revealed that his chapter still had "an open contract to kill Mick Jagger and the Rolling Stones" because of the legal flap after the 1969 Altamont concert incident.

"During the mid 1970s, our people sent a member with a .22 pistol and silencer to a hotel where the Stones were to stay. But it fell through because the band never showed up," he said.

Asked if his group used silenced weapons often, Butch replied, "Sure, what do you think would be more effective? Silencers are great for shooting at people."

What is truly rewarding about the American way is that Butch will find a useful occupation when his prison term is finished. There are several government agencies that can always use another employee with his talents and value.

Popular press stories in the early 1980s sensationalized how the KKK was heavily reorganizing and rearming. The Klan has included a variety of silenced weapons including M16s, Ingrams, Armalite rifles, and Ruger pistols. I read one police report in which an undercover informant who penetrated a Klan group says they had an American 180 with a "very professionally made and finely tuned sound moderator." At least one of the silenced Ruger pistols was documented in the Klan hit when member Hal Burdick was killed in California late in 1980.

And so it goes in Badguy Land, with the sound of gunfire silence broken only by the thuds of falling bodies.

11. Into Every Book Some Miscellany Must Fall

Even if you've won the Super Bowl, found an honest politician, or are living with an "11," nothing in the real world ties up neatly. That's why there is this chapter—my odds and ends drawer, my gold box full of miscellaneous tools and other stuff. This is the chapter which deals with the miscellany in the silencer kingdom—the politics, the philosophy, the construction, the use and, perhaps, the abuse.

Nearly three-quarters of a century before this book was written, a very brilliant, realistic, and outspoken arms inventor spoke out against the "Ordnance Ring" in the American defense establishment. He was Col. Isaac N. Lewis, inventor of myriad items military and civilian, but most famous for the excellent light machine gun that carries his name.

Embroiled in a political controversy replete with rigged ordnance trials and suspect buddy-buddy relationships between ordnance officials and major arms manufacturers, Isaac Lewis did something sinful: he told the truth to the American press. As a result, the Washington establishment blacklisted him, so he was forced to take his gun to Europe for adoption. The result of that episode is a sad bit of American military history. But this bit of history illustrates the problem still facing inventors, designers, and small producers of firearms: if they don't belong to "The Club" their designs and inventions are not bought. Sad and simple.

A case in point is the M16 rifle. Eugene Stoner is the real father of the M16. He is a brilliant man and his basic design reflects that. But Eugene Stoner couldn't sell his weapon, at the time called the AR15, to the defense establishment. Meanwhile, the social whirl of Washington waltzed by, with the major defense contractors and the Pentagon decision-makers all climbing into their cocktail party bed together.

Out came the redesigned bastard known as the M16. The names on the corporate birth certificate are brahmins of the ordnance industry. Therefore,

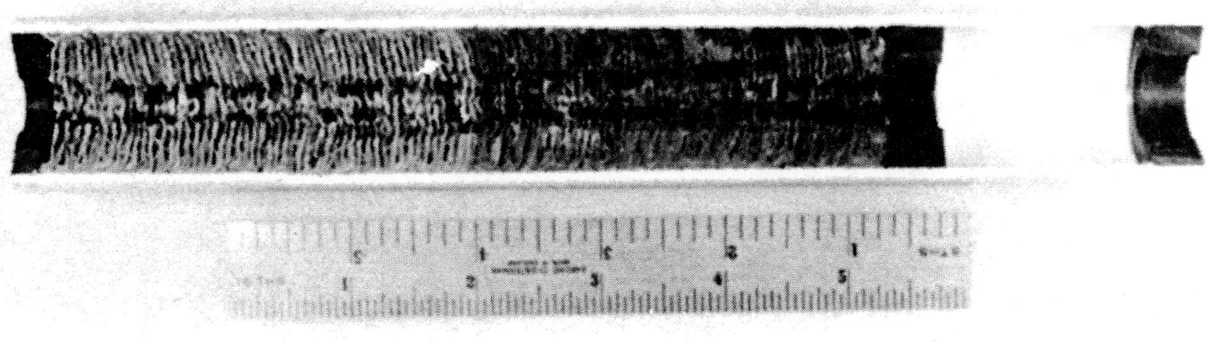

BILL WILSON

Home fabricated silencer is comprised of PVC water pipe and corrugated cardboard. Its designer says it "works and works very well."

JOHN A MINNERY

Troops of the American 28th Infantry of the 1st Division roll from their trenches late in May of 1918. Note soldier in lower center; his Springfield has a Maxim silencer mounted in place.

DONALD THOMAS

An early Ingram with a prototype two-stage suppressor.

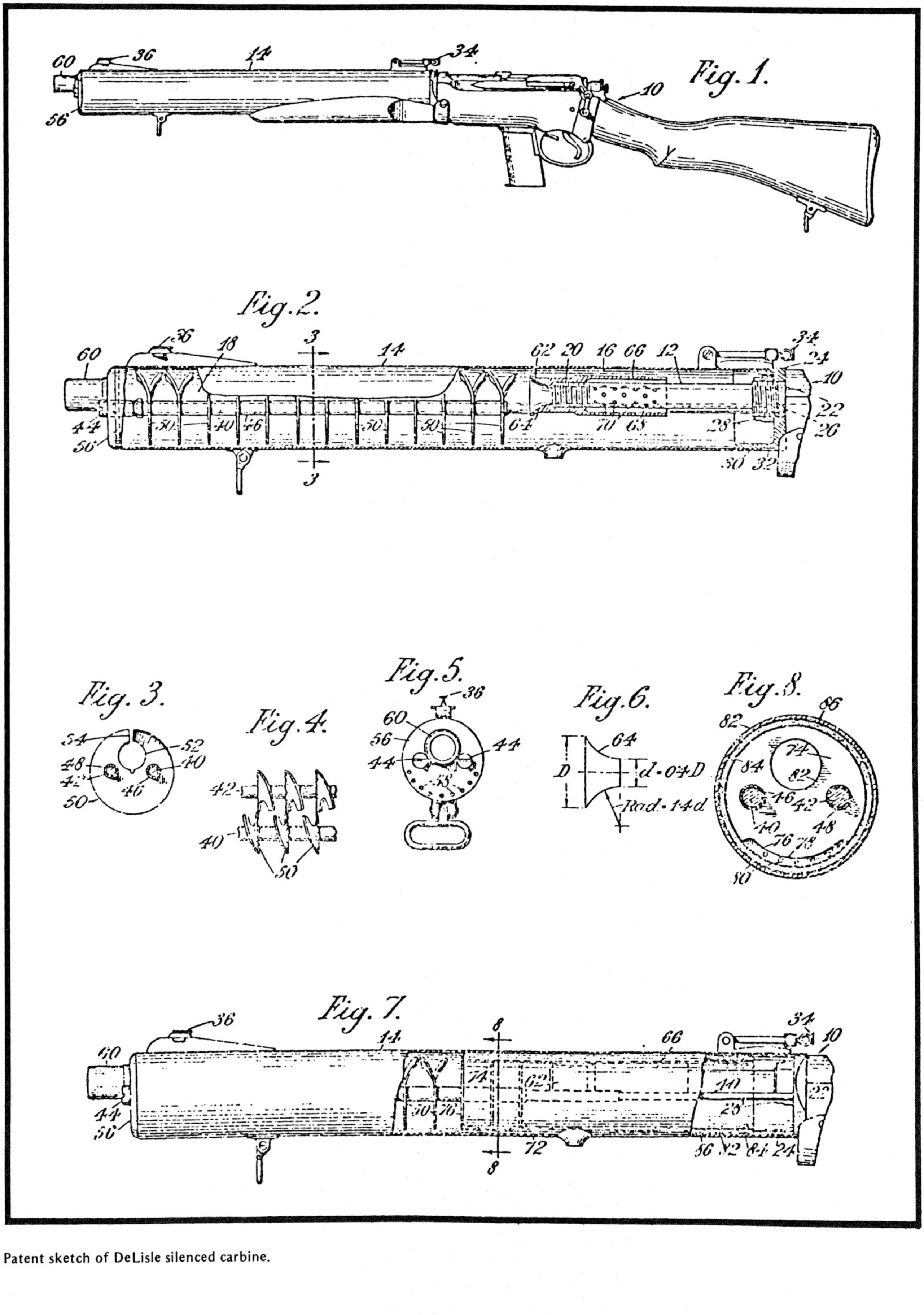

Patent sketch of DeLisle silenced carbine.

DICK SWANN

Louis Lavender test firing a Sionics suppressor on an M6A1 Lavender was on the Colt firearms range, working on a classified project.

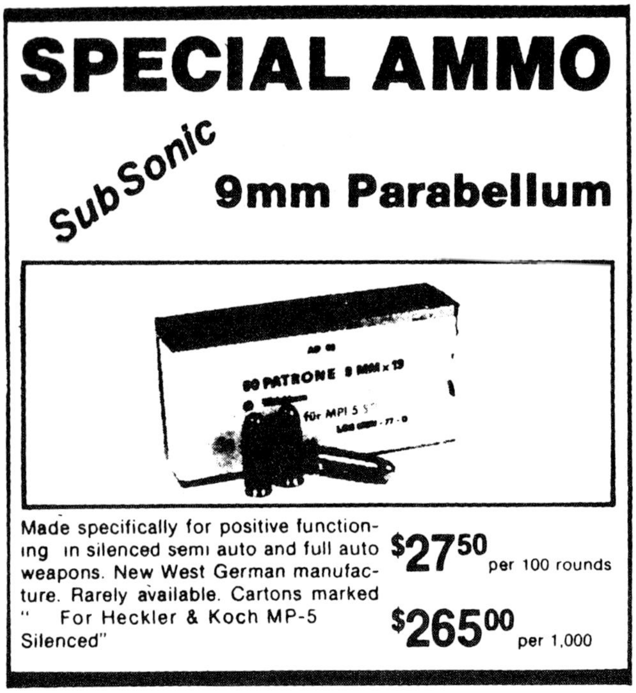

SHOTGUN NEWS/PMC

California's PMC began selling commercially loaded subsonic ammo for silenced weapons late in 1982.

it has to be good. Forget Eugene Stoner. Forget the Vietvets who had to use it. Remember: lobby, politics, money, powerful politicians, the sons and grandsons of the old "Ordnance Ring" Isaac Lewis exposed years ago. They're alive, well, and still making millions of bucks through their old boy banklines.

What is the kicker to all this? JSSAP decided to improve the M16A1. Just think: a wonderful way to retool, recontract, and reinsert the contractor's beak onto the public treasury teat to suck up all those defense dollars the Reaganistas have stashed in the Pentagon coffers. It sounds good. The ordnance corporations improve the rifle and make some millions doing it. Here's the irony: the "new improvements" are mostly the very same things Eugene Stoner had on his original model.

Sadly, this ordnance ring also goes through the nose of suppressor manufacturers. Only here, the healthy light of public sunshine does not get through the layers of secrecy and national embarrassment because silencers carry a classified stigma of shame, of something wrong. The design, development, production, and use of silencers is always classified because silenced weapons are generally used on missions that are legitimately dangerous and dirty, or they are used for political murder, which we officially disavow. The national myth is that the United States would never stoop to something so inhumane and dirty as political assassination. Yet the facts are plain. We have engaged in political assassination lots of times.

Beyond the heyday of Hiram Maxim, what we do read of the suppressor industry starts with Mitchell WerBell III and his ill-fated benchmarks of the past twenty years. The corporate demise of Mitch WerBell's suppressor-producing companies had one basic cause: the military establishment didn't see much of an officer's and a gentleman's need for suppressors, and, if it did, the suppressors would have to be purchased from one of the old school chaps already in the club—Isaac Lewis's "Ordnance Ring" connection. Thus, WerBell and companies were flushed away financially.

A few years of suppressor doldrums followed as peace broke out in Southeast Asia. Then, a new company grew from the WerBell wreckage. R. P. B. Industries, Inc., had as its principals Wayne Daniels, Robert Morgan, and a chap named LeBoult. Basically, R. P. B. tried to carry on in the WerBell line, producing Ingram M10 and M11

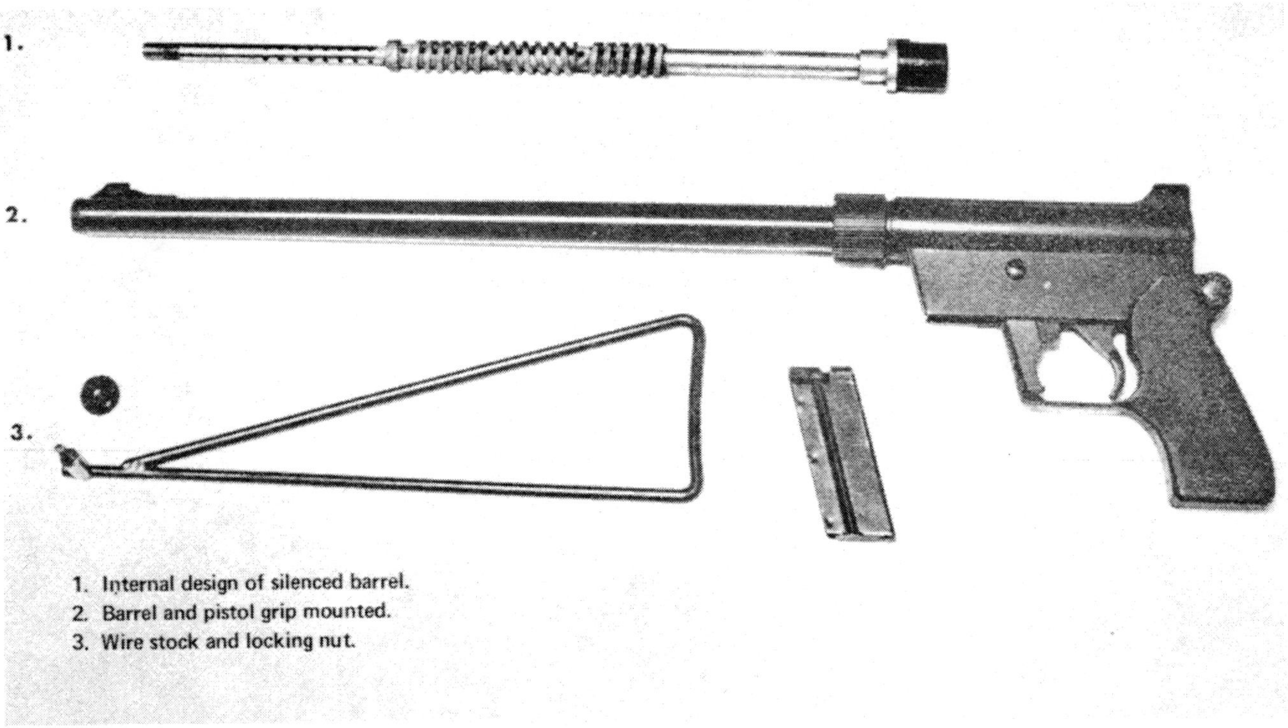

R. P. B. INDUSTRIES, INC.

A product literature photo showing parts for the R. P. B. suppressed AR7 survival rifle.

INTERDYNAMICS

The Interdynamics suppressor, special magazine and the subsonic ammunition they sell as a total suppression system.

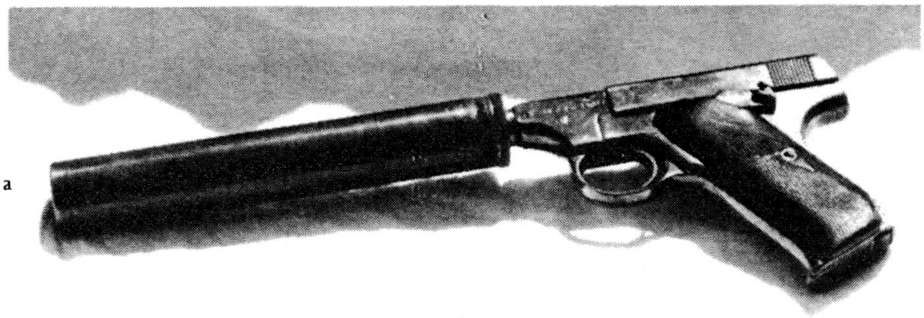

The .22 caliber sealed suppressor on a Colt Woodsman.

R. K THOMAS

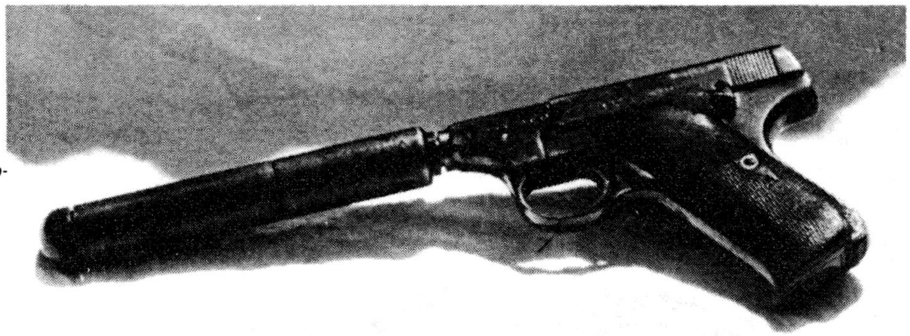

Colt Woodsman with .22 Gold Spot suppressor.

R K THOMAS

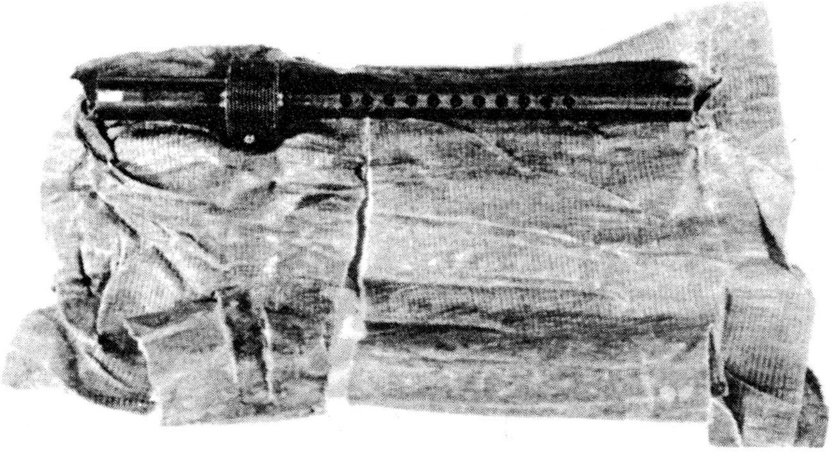

An original CIA SMG barrel, ported for suppressor use—possibly for a S & W M76 or an M3A1. The waterproofed wrapping is shown too.

G. GORDON DULLES

INTERDYNAMICS

Interdynamics special subsonic .223 load shown in full configuration and in cutaway (two views).

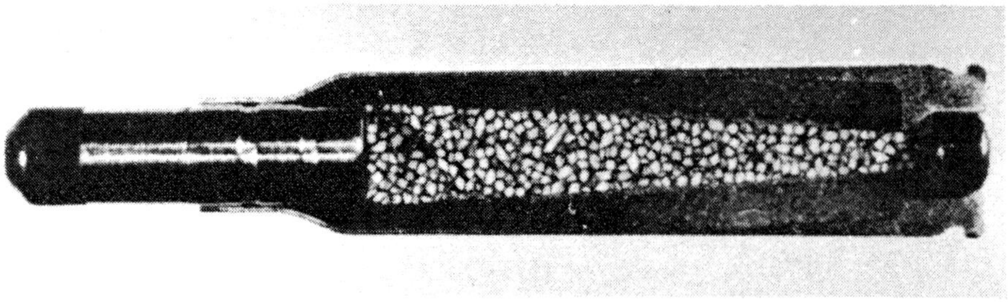

INTERDYNAMICS

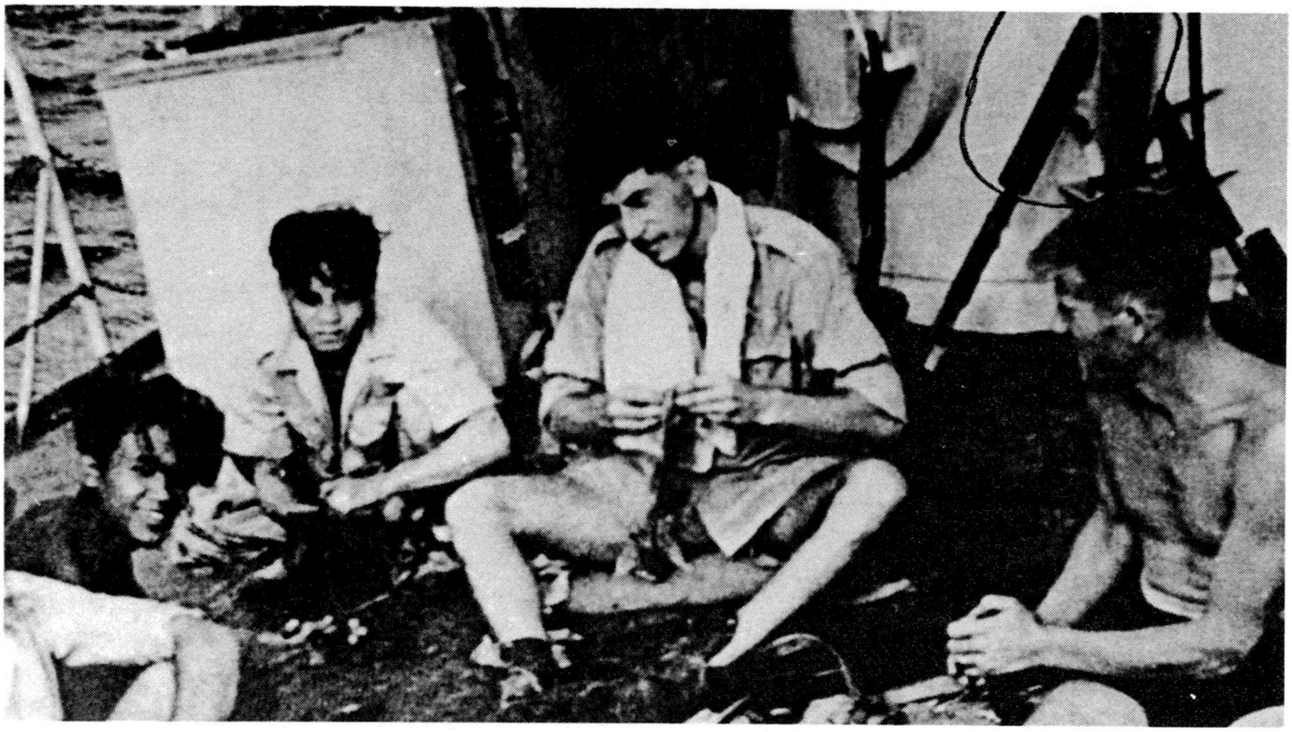
D. B. COOPER

This group of French counter-guerrilla fighters, circa 1951, has at least one silenced submachine gun stowed aboard for use in Indochina.

weapons, suppressors and accessories. Operating from a factory in Atlanta, and using much of the original equipment, they turned out an impressive array of products. They also sold the equipment, both to governments and individuals.

An R. P. B. ad exclaimed, "The Ingram SMG has gained world acclaim for awesome but reliable fire power. . . . In the summer of 1979, the Colombian government conducted extensive comparison tests on six of the best known submachine guns of the world, including the Ingram. The conclusion of these tests resulted in a large order from the Colombian government for Ingram guns."

In addition, other sales were made to the governments of England, Dubai, the Netherlands, Israel, Mexico, Brazil, Guatemala, Honduras, Portugal, Oman, Korea, and, of course, to the U.S. for the Army's Special Forces, the Navy SEALS, the Secret Service, and the FBI. I'm sure the CIA also bought a few which it has tucked into deep cover closets all over the globe.

According to Wayne Daniel, president of R. P. B., early in 1981 the company received an order from the Defense Department to produce fifty Ingram submachine guns equipped with suppressors for SEAL operations.

Squeezed by the ordnance ring and the fact that the military wasn't going to buy large numbers of M10 and M11 submachine guns, plus the trickle down effect that our allies weren't in the market either, R. P. B. put its engineers to work making semiautomatic versions of the Ingram guns for domestic consumer sales. As some buyers soon discovered, it was as relatively easy to convert the semiauto Ingrams back to full auto as it was to make them one-squeeze, one-shot weapons. Soon, other ads in *Shotgun News* offered ready-to-install conversion kits.

On 21 June 1982, the ATF declared that the semiauto versions of the Ingrams were too easy to convert to full auto and therefore were to be reclassified. The P/L statement didn't respond well to the ATF decision. So, in an effort to bolster sales, R. P. B. began offering suppressor kits early in 1982. They offered all internal parts for the following weapons at ninety dollars: Ingram M10 and M11, M16/AR15, M14/M1A1, AR7 and the Colt Woodsman. They also offered complete suppressors, which meant a numbered outer tube was added to the kit and the unit assembled for Class III dealers, police departments, and other legal purchasers.

However, it didn't work out, and on 18 October 1982, R. P. B. held a full public auction, selling its building, equipment, stock, and supplies right down to the janitor's brooms and mouse

R.K. THOMAS

Side view of the .22 caliber Woodsman with the Gold Spot suppressor in place.

R K THOMAS

Side view of the Colt Woodsman, .22, with a homebuilt suppressor from Australia, where possession of such units is quasi-illegal. Few arrests are made, however.

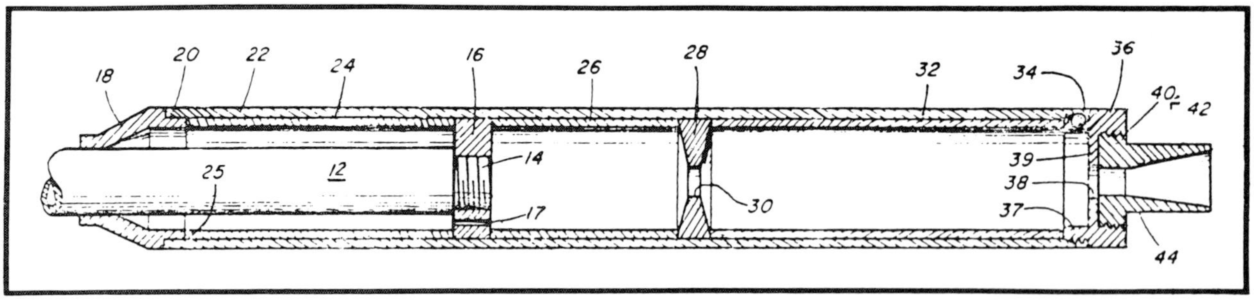

U S PATENT OFFICE

This unusual two-stage suppressor was designed for the U.S. Navy, reportedly for either the M16 or the CAR weapons carried by SEALs. It was tapered and closed with the chambers and a single internal baffle. No test data were available at the time of filing in 1969.

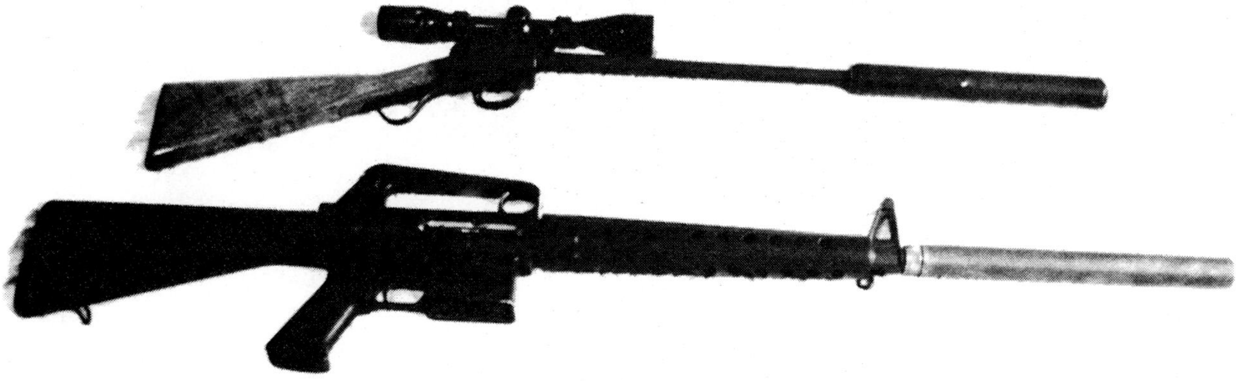

R. K. THOMAS

The longer Martini with homebuilt suppressor (top); an AR10 with another homemade suppressor (bottom). Both units tested excellently.

MARTIN CALLAHAN

Martin Callahan with a suppressor-equipped CAR.

traps. All the Ingrams, all the suppressor parts, all the machines that produced them, the ancillary equipment, the building, the parking lot—everything was sold. R. P. B. is no more, a victim of the economy, ATF harassment, and internal squabbling. The tooling and equipment was scattered all over the country as individuals bought various lots at the auction. It's certain that no one company will ever produce the weapons on an exclusive basis again.

According to some sources in suppressordom, at least one of the R. F. P. principals later surfaced as S.W.D., Inc., a survival-oriented gun supply shop in Atlanta. First advertising in 1983, they used both the Ingram and Cobray trademarks/logos in their ads, exclaiming that they believe in "survival of a quality product in great demand; the Ingram/Sionics type sound suppressor" parts sets.

Their introductory offer for internal suppressor parts was seventy-five dollars which did not include the outer tube. In addition, they advertised all parts for various Ingram weapons, plus registered conversions of many popular semiautomatics to

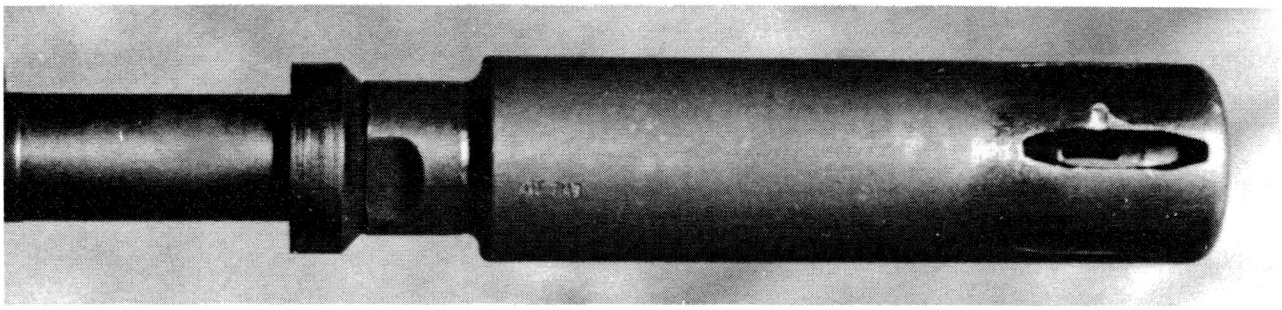

MARTIN CALLAHAN

Closeup of the suppressor on Callahan's CAR. Note ATF serial.

fully auto versions.

While terminal problems slowly strangled R. P. B., a successor was doing his incorporation homework. Shortly into 1982, Robert Miller began to publicize his Frankford Arsenal in Ft. Lauderdale, obviously trading on the name of the defunct federal facility in Philadelphia. In addition to his regular automatic weapons products, Miller had an interesting suppressor design for both submachine gun and rifle. According to Miller, his suppressor would be of the sealed type favored by Maxim.

"This design would be nonrebuildable, as the various styles of baffles would be welded together in a jig and then permanently sealed in the completed suppressor. No cleaning would be necessary, and there would be no need for a replaceable end wipe assembly," Miller notes.

In addition, he is manufacturing the standard Sionic-type suppressor, but claims a better performance than the original. He adds, "We have been able to achieve a lower decibel drop. I attribute this to holding closer tolerances on all machined and bearing surfaces, then simply taking time to see the work is done properly."

Miller's personal background also explains his philosophy. He says, "I come from a tool and die background, where far superior skills are required. We take great pride in our work here, and don't worry so much about making a million bucks. Our satisfaction comes from our finished product."

There were others. In 1980, *Ground Defence International* carried an announcement of a new Swedish suppressor system known as Interdynamic. The system consisted of a silencer and subsonic cartridge in 5.56mm. According to Interdynamic product literature, the system is designed for low-cost use in clandestine military and police operations.

The company modified the interior ballistics of the standard 5.56mm round, then altered its aerodynamic properties. The suppressor is what the product literature calls "a refinement of the well-proven Maxim multiple baffle type." With this combination of Interdynamic suppressor and subsonic ammunition, semiautomatic weapons must be manually operated, which is really not a disadvantage in most operations where quiet is vital. Interdynamic markets the system in the U.S. through an office in Florida.

Various companies have loaded and sold subsonic and other specially manufactured ammunition for silenced weapons. Probably the most proficient and top-quality work was done by Lee Juras with his late Super Vel Company. Occasionally, one sees ads for this type of ammunition in *Shotgun News,* plus the military orders small runs from its contractors. However, late in 1982, Pacific International Merchandising Corporation of Sacramento, California began marketing subsonic 9mm ammunition specifically for suppressed weapons. They advertised "positive functioning in silenced semiauto and full auto weapons. New, West German manufacture." The price was $265 per 1,000 rounds.

Australia has always been a busy silencer country and thanks to the help of R. K. Thomas, a Canadian who recently visited that country, I can offer some new information. Despite many of the other freedoms in Australia, silencers are not freely allowed. According to Thomas, ownership of silencers in most Australian states is illegal, while in others, possession is legal as long as the unit is not attached to a firearm.

However, he also related that experimentation and use of silencers there is ongoing. In one letter, Thomas notes, "They were experimenting with silenced Mauser in .30/06 and a Martini in .357. The Mauser silencer is physically unwieldy but it does reduce the sound to that of a high velocity .22 rimfire. The Martini is much better at .38 and not too much louder in .357. Both would be excellent with subsonic ammunition."

The design of the Martini silencer is quite simple, according to Thomas. It consists of a very large expansion chamber utilizing baffle discs spacers which then form smaller chambers. The end cap is metallurgically designed to reflect the gases backward. Thomas notes that the designers told him that the splendid efficiency is due to the initial gas expansion chamber being matched to the barrel in terms of volume release.

Another unit Thomas saw was a Colt Woodsman in .22. It took an original, sealed suppressor, as well as a commercial Goldspot unit. Thomas also passes along the information that in 1982, the Australian version of their Special Forces/SAS/Commandoes adopted the HK MP5 and the silenced MP5SD submachine guns as standard weapons.

Another Australian inventor, Guy Fawkes, designed his own .223 design, then redesigned a Sionics tube for his M16. He also has done an

TOM DUNKIN

Master suppressor king Mitch WerBell in the field, Southeast Asia, 1968 (two views).

excellent job with an original suppressor for the .45 M1911A1, using a newly developed integral barrel bushing and muzzle adapter worked up from an Ingram M10 unit.

All units work very well. R. K. Thomas, who tested Fawkes' .223, reports, "The (.223) design uses convergent-divergent flow passages with expansion chambers and a separate barrel sleeve system."

One of the more tantalizing stories generated by my earlier volumes came from Adam Dinterfass, who is researching various elite military units. He told me about a California company known as the "Subsonic Research Lab," which took ChiCom SKS rifles recovered in Vietnam and converted them for clandestine missions by CIA and Special Forces personnel in the 1960s and 1970s. The conversions included new barrels which were ported and designed for integrated silencer units.

By the way, Americans were not the initial users of suppressor-equipped weapons against the Communists in Vietnam. Discussing the tactical use of silencers during the 1950s Indochina wars, Robert Elford's narrator commented in *Devil's Guard* about the use of snipers with silenced weapons providing route security during missions into indigenous territory, as follows:

> We selected our few but trusted guides. They had been truly loyal to us and we respected them highly. If we passed by some rice paddies, for instance, where a few dozen peasants were at work, Eisner would give the word: "Abwehrmannschaft abtreten!" and six of our sharpshooters would quietly drop into the roadside underbrush, carrying telescopic rifles with silencers attached—a formidable weapon against guerrillas. The column would march on as though nothing had happened. Sometimes, and as soon as the army was out of sight, some peasants would turn into armed

J. DAVID TRUBY

In the old MAC plant, Marietta, Georgia, in 1972, a final assembly line photo of the Ingram SMG.

R. K. THOMAS

View of ported and threaded Woodsman barrel with suppressor removed.

U.S. Troopers on patrol, one of whom has an M14 with a Starlight scope and Sionics suppressor, prepare to move out east of the Cambodian/Vietnamese border. These men are with the 25th Infantry Division.

EMILIO SANTANA

In the field with Mitch WerBell in 1968, demonstrating Sionac suppressors with the M16A1 in Southeast Asia.

MITCHELL WERBELL III

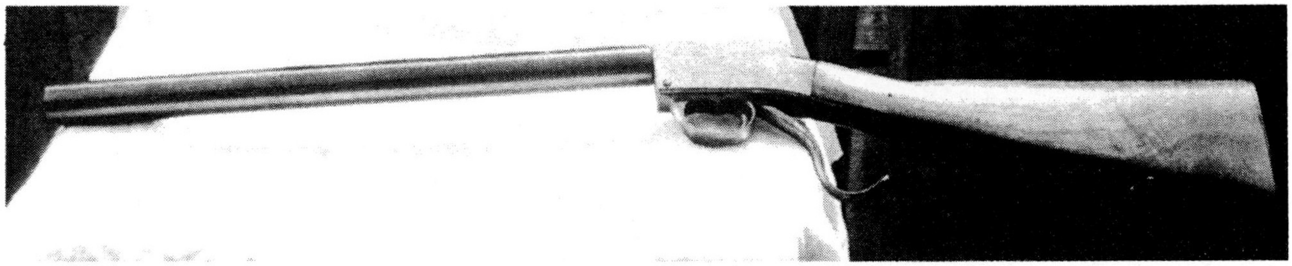

R. K. THOMAS

The Canadian R. K. Thomas tested this silenced Martini rifle in .357 and found it an excellent unit. The suppressor is an integral model.

terrorists, taking off after the column head over heels. Our sharpshooters would drop them before they reached the jungle.

It was also one of our tricks to pass a Viet Minh-controlled village without bothering a soul. The column would vanish into the hills, except for the sharpshooters, who would drop back to cover every exit. In ninety percent of all cases, Viet Minh messengers or even groups of guerrillas would emerge from the village and depart in a hurry. The silencer-equipped guns were excellent for dropping them quickly and quietly. Indeed, our marksmen were capable of hitting a dozen terrorists within a few seconds, starting invariably with the last man in a line or group. Erich Schulze had once eliminated five running guerrillas, repeating aloud, "Mitte-mitte-mitte-mitte-mitte"—"Center-center . . .," pulling the trigger at each word which corresponded with one shot per second. We had used the same ruse in occupied Russia and invariably it worked.

Even though the American forces are in very low profile in Asia these days, our weapons are not. The Nationalist Chinese Marines, part of the elite Amphibious Reconnaissance Patrol (ARP) Regiment, use silenced weapons on some of their special missions. Their standard weapon is the Ingram M10 with suppressor. You know, with as many units, countries and organizations as there seem to be credited with using the silenced Ingrams, one wonders why someone hasn't managed at least a modest fortune from that weapon. In law enforcement sales alone, both domestic and import, I'd figure financial success for the Ingram, despite lack of quality during some of the manufacturing eras.

While the basic requirements for military suppressors are both obvious and documented in great detail, the law enforcement field is another arena where the reduction of firearms noise has many benefits.

For years, police viewed silencers as the tools of spies, assassins or those likely to shoot their own mothers in the back. Today's law enforcement thinking, though, seems to be strongly prosup-

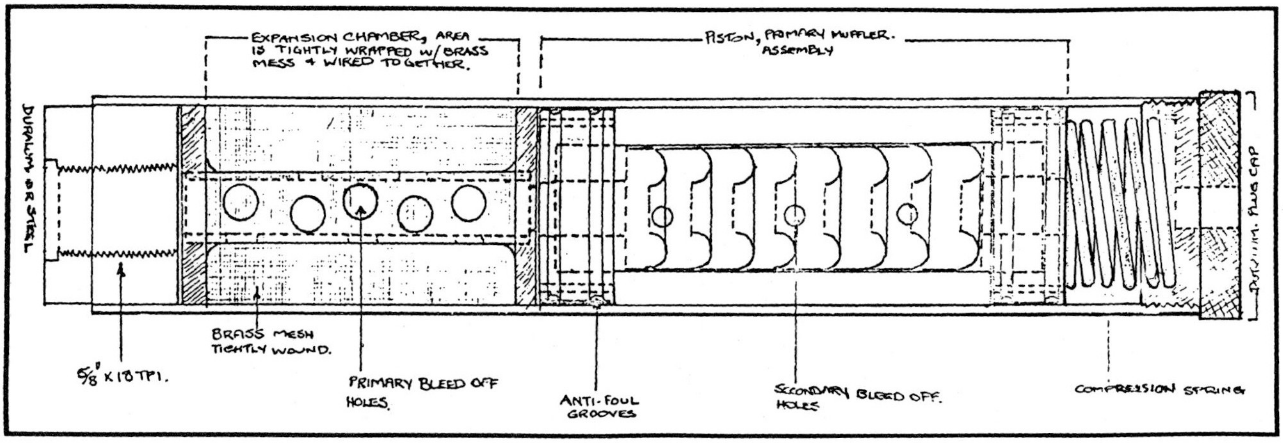

R. K. THOMAS

Schematic drawing done by Thomas to show the design of the Australian suppressor for the .30-06 Mauser, a unit of porting and baffles with brass mesh.

pressor. Taking the lead in technology, the FBI has adopted a large number of suppressed weapons appropriate to law enforcement use including the Uzi and HK MP5SD submachine guns, the M16A1, XM21, and M40 rifles, plus a number of handguns, including Ruger and High Standard in .22 caliber, S & W, Beretta and 9mm HK, plus a variety of other types.

The obvious uses for suppressed weapons include taking out a hostage-holder with the quiet precision of a S.W.A.T. team sniper rifle at 200 meters, quietly neutralizing a large narcotics operation's guard dogs, and avoiding public panic in critical shootout situations unfolding in crowded urban areas. Then, there is the executive protection business, where a suppressor on security personnel's compact, powerful weapons will prevent deafness or concussion from the awesome muzzle blast inside the confined space of an automobile or elevator. As one who has fired an Ingram M10 inside a vehicle, my ears and senses were thankful for the suppressor on the weapon.

From a public relations standpoint, the lowered noise level from a suppressor-equipped weapon will minimize critical civilian reaction to gunfire. These types of weapons also confuse a sniper or hostage-taker as he or she is not really certain from where police gunfire is coming. Finally, they are ideal weapons for dealing with injured, dangerous, or wild animals that must be destroyed in an urban or suburban area.

In addition to military and law enforcement use, collectors and others who enjoy exotic weaponry are buying suppressors and suppressed weapons in record numbers, according to dealers I spoke with. That, of course, means prices are up. As one dealer related to me, "Prices continue to grow like weeds in a manure pile."

A 1982 classified ad in *Shotgun News* listed an original Sionics M14 suppressor for $1,025, an OSS High Standard suppressed pistol for $1,300, and a Sten MkIIS for $1,500. In a more modern view, preauction MAC 9mm suppressors were going for $185 in the fall of 1982. By the first month of 1983, the LARAND M16 suppressor cost $425, the LARAND Ruger MkI cost $520, and a High Standard HD with a LARAND unit cost $610.

One of the greatest collections of militaria was sold for millions when the York Arms & Armor Museum collection was sold at auction in Las Vegas in December of 1981. Some of the rarest weapons in the world changed hands during this extraordinary sale, including collector's specimens of prime suppressors. Even original Maxim, Sionics, MAC, Colt, Sterling, and Sten suppressors and suppressed weapons were included.

Silencers may not be the gun world's better financial mousetrap, but they surely do take a lot of the SNAP out of getting caught.

12. Sources . . .

Compiled by J. David Truby,
Feb. 1983

American Ballistics Co.
P.O. Box 1410
Marietta, GA 30061
(404) 434-8087

Automatic Weapons Co.
P.O. Box 1731
Socorro, NM 87801

Brown & Associates
7370 Twin Branch Road, NE
Atlanta, GA 30328
(404) 394-3701

C & C Tubing
Box 50019
Cicero, IL 60650

R. E. Choate
1904 11th Avenue
Yuma, AZ 85364
(602) 783-8624

Jonathan Arthur Ciener
1115 N. Courtenay Parkway
Merritt Island, FL 32952

Craig Improved Armaments, Inc.
P.O. Box 595
Coushatta, LA 71019

D. A. Q.
Box 50269
Cicero, IL 60650

George Dodson
Box 09161
Detroit, MI 48209

Frankford Arsenal
1047 NE 43rd Ct.
Ft. Lauderdale, FL 33334
(305) 566-8690

Interdynamic of America, Inc.
223 SW 22 Avenue
Miami, FL 33135

INTERRAND
1941 S. Arlington Ridge Road
Arlington, VA 22202

MAC
Caesars Head Road
Cleveland, SC 29635
(803) 836-6326

New Frontier Armament Co.
2347 N. Edgewood Avenue
Jacksonville, FL 32205
(904) 781-0509

R. F. P.
P.O. Box 971
Milpitas, CA 95035
(408) 946-0677

SIONAC
P.O. Box 5624
Tucson, AZ 85703
(602) 888-7490

Special Missions Group, Ltd.
P.O. Box 254
Soda Springs, CA 95728

J. R. Stemple
4716 Bentham Drive
Columbus, OH 43220
(614) 457-8433

Survival Supply Co.
P.O. Box 372
Mountain Home, AR 72653

S. W. A.
P.O. Box 12017
El Casom, CA 92022

Please always include a self-addressed, stamped envelope with any inquiry.